The Selfish Gene

Richard Dawkins

THE SELFISH GENE

OXFORD UNIVERSITY PRESS

1976

Oxford University Press, Walton Street, Oxford OX2 6DP

OXFORD LONDON GLASGOW NEW YORK
TORONTO MELBOURNE WELLINGTON CAPE TOWN
IBADAN NAIROBI DAR ES SALAAM LUSAKA ADDIS ABABA
KUALA LUMPUR SINGAPORE JAKARTA HONG KONG TOKYO
DELHI BOMBAY CALCUTTA MADRAS KARACHI

© OXFORD UNIVERSITY PRESS 1976

British Library Cataloguing in Publication Data
Dawkins, Richard
  The selfish gene.
  ISBN 0–19–857519–X
  1. Title
  591.5   QL751
  Animals, Habits and behaviour of
  Animals – Genetics

FILMSET IN 'MONOPHOTO' EHRHARDT 11 ON 12 PT BY
RICHARD CLAY (THE CHAUCER PRESS), LTD., BUNGAY, SUFFOLK
AND PRINTED IN GREAT BRITAIN BY
FLETCHER & SON LTD., NORWICH

Foreword

THE chimpanzee and the human share about 99·5 per cent of their evolutionary history, yet most human thinkers regard the chimp as a malformed, irrelevant oddity while seeing themselves as stepping-stones to the Almighty. To an evolutionist this cannot be so. There exists no objective basis on which to elevate one species above another. Chimp and human, lizard and fungus, we have all evolved over some three billion years by a process known as natural selection. Within each species some individuals leave more surviving offspring than others, so that the inheritable traits (genes) of the reproductively successful become more numerous in the next generation. This is natural selection: the non-random differential reproduction of genes. Natural selection has built us, and it is natural selection we must understand if we are to comprehend our own identities.

Although Darwin's theory of evolution through natural selection is central to the study of social behavior (especially when wedded to Mendel's genetics), it has been very widely neglected. Whole industries have grown up in the social sciences dedicated to the construction of a pre-Darwinian and pre-Mendelian view of the social and psychological world. Even within biology the neglect and misuse of Darwinian theory has been astonishing. Whatever the reasons for this strange development, there is evidence that it is coming to an end. The great work of Darwin and Mendel has been extended by a growing number of workers, most notably by R. A. Fisher, W. D. Hamilton, G. C. Williams, and J. Maynard Smith. Now, for the first time, this important body of social theory based on natural selection is presented in a simple and popular form by Richard Dawkins.

One by one Dawkins takes up the major themes of the new work in social theory: the concepts of altruistic and selfish behavior, the genetical definition of self-interest, the evolution of aggressive behavior, kinship theory (including parent–offspring relations and the evolution of the social insects), sex ratio theory, reciprocal altruism, deceit, and the natural selection of sex differences. With a confidence that comes from mastering the underlying theory, Dawkins unfolds the new work with admirable clarity and style. Broadly educated in biology, he gives the reader a taste of its rich and fascinating literature. Where he differs from published work (as he does in criticizing a fallacy of my own), he is almost invariably exactly on target. Dawkins also takes pains to make clear the logic of his arguments, so that the reader, by applying the logic given, can extend the arguments (and even take on Dawkins himself). The arguments themselves extend in many directions. For example, if (as Dawkins argues) deceit is fundamental to animal communication, then there must be strong selection to spot deception and this ought, in turn, to select for a degree of self-deception, rendering some facts and motives unconscious so as not to betray—by the subtle signs of self-knowledge—the deception being practised. Thus, the conventional view that natural selection favors nervous systems which produce ever more accurate images of the world must be a very naïve view of mental evolution.

The recent progress in social theory has been substantial enough to have generated a minor flurry of counter-revolutionary activity. It has been alleged, for example, that the recent progress is, in fact, part of a cyclical conspiracy to impede social advancement by making such advancement appear to be genetically impossible. Similar feeble thoughts have been strung together to produce the impression that Darwinian social theory is reactionary in its political implications. This is very far from the truth. The genetic equality of the sexes is, for the first time, clearly established by Fisher and Hamilton. Theory and quantitative data from the social insects demonstrate that there is no inherent tendency for parents to dominate their offspring (or vice versa). And the concepts of parental investment and female choice provide an objective and unbiased basis for viewing sex differences, a considerable advance over popular efforts to root women's powers and rights in the functionless swamp of biological identity. In short, Darwinian social theory gives us a glimpse of an underlying symmetry and logic in social relationships which, when more fully comprehended by ourselves, should revitalize our political understanding and provide the intellectual support for a science and medicine of psychology. In the process it should also give us a deeper understanding of the many roots of our suffering.

ROBERT L. TRIVERS

Harvard University
July, 1976

Preface

THIS book should be read almost as though it were science fiction. It is designed to appeal to the imagination. But it is not science fiction: it is science. Cliché or not, 'stranger than fiction' expresses exactly how I feel about the truth. We are survival machines—robot vehicles blindly programmed to preserve the selfish molecules known as genes. This is a truth which still fills me with astonishment. Though I have known it for years, I never seem to get fully used to it. One of my hopes is that I may have some success in astonishing others.

Three imaginary readers looked over my shoulder while I was writing, and I now dedicate the book to them. First the general reader, the layman. For him I have avoided technical jargon almost totally, and where I have had to use specialized words I have defined them. I now wonder why we don't censor most of our jargon from learned journals too. I have assumed that the layman has no special knowledge, but I have not assumed that he is stupid. Anyone can popularize science if he oversimplifies. I have worked hard to try to popularize some subtle and complicated ideas in non-mathematical language, without losing their essence. I do not know how far I have succeeded in this, nor how far I have succeeded in another of my ambitions: to try to make the book as entertaining and gripping as its subject matter deserves. I have long felt that biology ought to seem as exciting as a mystery story, for a mystery story is exactly what biology is. I do not dare to hope that I have conveyed more than a tiny fraction of the excitement which the subject has to offer.

My second imaginary reader was the expert. He has been a harsh critic, sharply drawing in his breath at some of my analogies and figures of speech. His favourite phrases are 'with the exception of'; 'but on the other hand'; and 'ugh'. I listened to him attentively, and even completely rewrote one chapter entirely for his benefit, but in the end I have had to tell the story my way.

The expert will still not be totally happy with the way I put things. Yet my greatest hope is that even he will find something new here; a new way of looking at familiar ideas perhaps; even stimulation of new ideas of his own. If this is too high an aspiration, may I at least hope that the book will entertain him on a train?

The third reader I had in mind was the student, making the transition from layman to expert. If he still has not made up his mind what field he wants to be an expert in, I hope to encourage him to give my own field of zoology a second glance. There is a better reason for studying zoology than its possible 'usefulness', and the general likeableness of animals. This reason is that we animals are the most complicated and perfectly-designed pieces of machinery in the known universe. Put it like that, and it is hard to see why anybody studies anything else! For the student who has already committed himself to zoology, I hope my book may have some educational value. He is having to work through the original papers and technical books on which my treatment is based. If he finds the original sources hard to digest, perhaps my non-mathematical interpretation may help, as an introduction and adjunct.

There are obvious dangers in trying to appeal to three different kinds of reader. I can only say that I have been very conscious of these dangers, but that they seemed to be outweighed by the advantages of the attempt.

I am an ethologist, and this is a book about animal behaviour. My debt to the ethological tradition in which I was trained will be obvious. In particular, Niko Tinbergen does not realize the extent of his influence on me during the twelve years I worked under him at Oxford. The phrase 'survival machine', though not actually his own, might well be. But ethology has recently been invigorated by an invasion of fresh ideas from sources not conventionally regarded as ethological. This book is largely based on these new ideas. Their originators are acknowledged in the appropriate places in the text; the dominant figures are G. C. Williams, J. Maynard Smith, W. D. Hamilton, and R. L. Trivers.

Various people suggested titles for the book, which I have gratefully used as chapter titles: 'Immortal Coils', John Krebs; 'The Gene Machine', Desmond Morris; 'Genesmanship', Tim

Clutton-Brock and Jean Dawkins, independently with apologies to Stephen Potter.

Imaginary readers may serve as targets for pious hopes and aspirations, but they are of less practical use than real readers and critics. I am addicted to revising, and Marian Dawkins has been subjected to countless drafts and redrafts of every page. Her considerable knowledge of the biological literature and her understanding of theoretical issues, together with her ceaseless encouragement and moral support, have been essential to me. John Krebs too read the whole book in draft. He knows more about the subject than I do, and he has been generous and unstinting with his advice and suggestions. Glenys Thomson and Walter Bodmer criticized my handling of genetic topics kindly but firmly. I fear that my revision may still not fully satisfy them, but I hope they will find it somewhat improved. I am most grateful for their time and patience. John Dawkins exercised an unerring eye for misleading phraseology, and made excellent constructive suggestions for re-wording. I could not have wished for a more suitable 'intelligent layman' than Maxwell Stamp. His perceptive spotting of an important general flaw in the style of the first draft did much for the final version. Others who constructively criticized particular chapters, or otherwise gave expert advice, were John Maynard Smith, Desmond Morris, Tom Maschler, Nick Blurton Jones, Sarah Kettlewell, Nick Humphrey, Tim Clutton-Brock, Louise Johnson, Christopher Graham, Geoff Parker, and Robert Trivers. Pat Searle and Stephanie Verhoeven not only typed with skill, but encouraged me by seeming to do so with enjoyment. Finally, I wish to thank Michael Rodgers of Oxford University Press who, in addition to helpfully criticizing the manuscript, worked far beyond the call of duty in attending to all aspects of the production of this book.

RICHARD DAWKINS

Contents

1. Why are people?

INTELLIGENT life on a planet comes of age when it first works out the reason for its own existence. If superior creatures from space ever visit earth, the first question they will ask, in order to assess the level of our civilization, is: 'Have they discovered evolution yet?' Living organisms had existed on earth, without ever knowing why, for over three thousand million years before the truth finally dawned on one of them. His name was Charles Darwin. To be fair, others had had inklings of the truth, but it was Darwin who first put together a coherent and tenable account of why we exist. Darwin made it possible for us to give a sensible answer to the curious child whose question heads this chapter. We no longer have to resort to superstition when faced with the deep problems: Is there a meaning to life? What are we for? What is man? After posing the last of these questions, the eminent zoologist G. G. Simpson put it thus: 'The point I want to make now is that all attempts to answer that question before 1859 are worthless and that we will be better off if we ignore them completely.'

Today the theory of evolution is about as much open to doubt as the theory that the earth goes round the sun, but the full implications of Darwin's revolution have yet to be widely realized. Zoology is still a minority subject in universities, and even those who choose to study it often make their decision without appreciating its profound philosophical significance. Philosophy and the subjects known as 'humanities' are still taught almost as if Darwin had never lived. No doubt this will change in time. In any case, this book is not intended as a general advocacy of Darwinism. Instead, it will explore the consequences of the evolution theory for a particular issue. My purpose is to examine the biology of selfishness and altruism.

Apart from its academic interest, the human importance of this subject is obvious. It touches every aspect of our social lives, our loving and hating, fighting and cooperating, giving and stealing, our greed and our generosity. These are claims which could have been made for Lorenz's *On Aggression*, Ardrey's *The Social Contract*, and Eibl-Eibesfeldt's *Love and Hate*. The trouble with these books is that their authors got it totally and utterly wrong. They got it wrong because they misunderstood how evolution works. They made the erroneous assumption that the important thing in evolution is the good of the *species* (or the group) rather than the good of the individual (or the gene). It is ironic that Ashley Montagu should criticize Lorenz as a 'direct descendant of the "nature red in tooth and claw" thinkers of the nineteenth century . . .'. As I understand Lorenz's view of evolution, he would be very much at one with Montagu in rejecting the implications of Tennyson's famous phrase. Unlike both of them, I think 'nature red in tooth and claw' sums up our modern understanding of natural selection admirably.

Before beginning on my argument itself, I want to explain briefly what sort of an argument it is, and what sort of an argument it is not. If we were told that a man had lived a long and prosperous life in the world of Chicago gangsters, we would be entitled to make some guesses as to the sort of man he was. We might expect that he would have qualities such as toughness, a quick trigger finger, and the ability to attract loyal friends. These would not be infallible deductions, but you can make some inferences about a man's character if you know something about the conditions in which he has survived and prospered. The argument of this book is that we, and all other animals, are machines created by our genes. Like successful Chicago gangsters, our genes have survived, in some cases for millions of years, in a highly competitive world. This entitles us to expect certain qualities in our genes. I shall argue that a predominant quality to be expected in a successful gene is ruthless selfishness. This gene selfishness will usually give rise to selfishness in individual behaviour. However, as we shall see, there are special circumstances in which a gene can achieve its own selfish goals best by fostering a limited form of altruism at the level of individual animals. 'Special' and 'limited' are important words in the last sentence. Much as we might wish to believe otherwise, universal love and the welfare of the species as a whole are concepts which simply do not make evolutionary sense.

This brings me to the first point I want to make about what this book is *not*. I am not advocating a morality based on evolution. I am saying how things have evolved. I am not saying how we humans morally ought to behave. I stress this, because I know I am in danger of being misunderstood by those people, all too numerous, who cannot distinguish a statement of belief in what is the case from an advocacy of what ought to be the case. My own feeling is that a human society based simply on the gene's law of universal ruthless selfishness would be a very nasty society in which to live. But unfortunately, however much we may deplore something, it does not stop it being true. This book is mainly intended to be interesting, but if you would extract a moral from it, read it as a warning. Be warned that if you wish, as I do, to build a society in which individuals cooperate generously and unselfishly towards a common good, you can expect little help from biological nature. Let us try to *teach* generosity and altruism, because we are born selfish. Let us understand what our own selfish genes are up to, because we may then at least have the chance to upset their designs, something which no other species has ever aspired to.

As a corollary to these remarks about teaching, it is a fallacy—incidentally a very common one—to suppose that genetically inherited traits are by definition fixed and unmodifiable. Our genes may instruct us to be selfish, but we are not necessarily compelled to obey them all our lives. It may just be more difficult to learn altruism that it would be if we were genetically programmed to be altruistic. Among animals, man is uniquely dominated by culture, by influences learned and handed down. Some would say that culture is so important that genes, whether selfish or not, are virtually irrelevant to the understanding of human nature. Others would disagree. It all depends where you stand in the debate over 'nature versus nurture' as determinants of human attributes. This brings me to the second thing this book is not: it is not an advocacy of one position or another in the nature/nurture controversy. Naturally I have an opinion on this, but I am not going to express it, except insofar as it is implicit in the view of culture which I shall present in the final chapter. If genes really turn out to be totally irrelevant to the determination of modern human behaviour, if we really are unique among animals in this respect, it is, at the very least, still interesting to inquire about the rule to which we have so recently become the exception. And if our species is not so exceptional as we might like to think, it is even more important that we should study the rule.

The third thing this book is not is a descriptive account of the detailed behaviour of man or of any other particular animal species. I shall use factual details only as illustrative examples. I shall not be saying: 'If you look at the behaviour of baboons you will find it to be selfish; therefore the chances are that human behaviour is selfish also'. The logic of my 'Chicago gangster' argument is quite different. It is this. Humans and baboons have evolved by natural selection. If you look at the way natural selection works, it seems to follow that anything that has evolved by natural selection should be selfish. Therefore we must expect that when we go and look at the behaviour of baboons, humans, and all other living creatures, we will find it to be selfish. If we find that our expectation is wrong, if we observe that human behaviour is truly altruistic, then we will be faced with something puzzling, something that needs explaining.

Before going any further, we need a definition. An entity, such as a baboon, it is said to be altruistic if it behaves in such a way as to increase another such entity's welfare at the expense of its own. Selfish behaviour has exactly the opposite effect. 'Welfare' is defined as 'chances of survival', even if the effect on actual life and death prospects is so small as to *seem* negligible. One of the surprising consequences of the modern version of the Darwinian theory is that apparently trivial tiny influences on survival probability can have a major impact on evolution. This is because of the enormous time available for such influences to make themselves felt.

It is important to realize that the above definitions of altruism and selfishness are *behavioural*, not subjective. I am not concerned here with the psychology of motives. I am not going to argue about whether people who behave altruistically are 'really' doing it for secret or subconscious selfish motives. Maybe they are and maybe they aren't, and maybe we can never know, but in any case that is not what this book is about. My definition is concerned only with whether the *effect* of an act is to lower or raise the survival prospects of the presumed altruist and the survival prospects of the presumed beneficiary.

It is a very complicated business to demonstrate the effects of behaviour on long-term survival prospects. In practice, when we apply the definition to real behaviour, we must qualify it with the word 'apparently'. An apparently altruistic act is one which looks, superficially, as if it must tend to make the altruist more likely (however slightly) to die, and the recipient more likely to survive. It often turns out on closer inspection that acts of apparent altruism are really selfishness in disguise. Once again, I do not mean that the underlying motives are secretly selfish, but that the real effects of the act on survival prospects are the reverse of what we originally thought.

I am going to give some examples of apparently selfish and apparently altruistic behaviour. It is difficult to suppress subjective habits of thought when we are dealing with our own species, so I shall choose examples from other animals instead. First some miscellaneous examples of selfish behaviour by individual animals.

Blackheaded gulls nest in large colonies, the nests being only a few feet apart. When the chicks first hatch out they are small and defenceless and easy to swallow. It is quite common for a gull to wait until a neighbour's back is turned, perhaps while it is away fishing, and then pounce on one of the neighbour's chicks and swallow it whole. It thereby obtains a good nutritious meal, without having to go to the trouble of catching a fish, and without having to leave its own nest unprotected.

More well known is the macabre cannibalism of female praying mantises. Mantises are large carnivorous insects. They normally eat smaller insects such as flies, but they will attack almost anything that moves. When they mate, the male cautiously creeps up on the female, mounts her, and copulates. If the female gets the chance, she will eat him, beginning by biting his head off, either as the male is approaching, or immediately after he mounts, or after they separate. It might seem most sensible for her to wait until copulation is over before she starts to eat him. But the loss of the head does not seem to throw the rest of the male's body off its sexual stride. Indeed, since the insect head is the seat of some inhibitory nerve centres, it is possible that the female improves the male's sexual performance by eating his head. If so, this is an added benefit. The primary one is that she obtains a good meal.

The word 'selfish' may seem an understatement for such extreme cases as cannibalism, although these fit well with our definition. Perhaps we can sympathize more directly with the reported cowardly behaviour of emperor penguins in the Antarctic. They have been seen standing on the brink of the water, hesitating before diving in, because of the danger of being eaten by seals. If only one of them would dive in, the rest would know whether there was a seal there or not. Naturally nobody wants to be the guinea pig, so they wait, and sometimes even try to push each other in.

More ordinarily, selfish behaviour may simply consist in refusing to share some valued resource such as food, territory, or sexual partners. Now for some examples of apparently altruistic behaviour.

The stinging behaviour of worker bees is a very effective defence against honey robbers. But the bees who do the stinging are kamikaze fighters. In the act of stinging, vital internal organs are usually torn out of the body, and the bee dies soon afterwards. Her suicide mission may have saved the colony's vital food stocks, but she herself is not around to reap the benefits. By our definition this is an altruistic behavioural act. Remember that we are not talking about conscious motives. They may or may not be present, both here and in the selfishness examples, but they are irrelevant to our definition.

Laying down one's life for one's friends is obviously altruistic, but so also is taking a slight risk for them. Many small birds, when they see a flying predator such as a hawk, give a characteristic 'alarm call', upon which the whole flock takes appropriate evasive action. There is indirect evidence that the bird who gives the alarm call puts itself in special danger, because it attracts the predator's attention particularly to itself. This is only a slight additional risk, but it nevertheless seems, at least at first sight, to qualify as an altruistic act by our definition.

The commonest and most conspicuous acts of animal altruism are done by parents, especially mothers, towards their children. They may incubate them, either in nests or in their own bodies, feed them at enormous cost to themselves, and take great risks in protecting them from predators. To take just one particular example, many ground-nesting birds perform a so-called 'distraction display' when a predator such as a fox approaches. The parent bird limps away from the nest, holding out one wing as though it were broken. The predator, sensing easy prey, is lured away from the nest containing the chicks. Finally the parent bird gives up its pretence and leaps into the air just in time to escape the fox's jaws. It has probably saved the life of its nestlings, but at some risk to itself.

I am not trying to make a point by telling stories. Chosen examples are never serious evidence for any worthwhile generalization. These stories are simply intended as illustrations of what I mean by altruistic and selfish behaviour at the level of individuals. This book will show how both individual selfishness and individual altruism are explained by the fundamental law which I am calling *gene selfishness*. But first I must deal with a particular erroneous explanation for altruism, because it is widely known, and even widely taught in schools.

This explanation is based on the misconception which I have already mentioned, that living creatures evolve to do things 'for the good of the species' or 'for the good of the group'. It is easy to see how this idea got its start in biology. Much of an animal's life is devoted to reproduction, and most of the acts of altruistic self-sacrifice which are observed in nature are performed by parents towards their young. 'Perpetuation of the species' is a common euphemism for reproduction, and it is undeniably a *consequence* of reproduction. It requires only a slight over-stretching of logic to deduce that the 'function' of reproduction is 'to' perpetuate the species. From this it is but a further short false step to conclude that animals will in general behave in such a way as to favour the perpetuation of the species. Altruism towards fellow members of the species seems to follow.

This line of thought can be put into vaguely Darwinian terms. Evolution works by natural selection, and natural selection means the differential survival of the 'fittest'. But are we talking about the fittest individuals, the fittest races, the fittest species, or what? For some purposes this does not greatly matter, but when we are talking about altruism it is obviously crucial. If it is species which are competing in what Darwin called the struggle for existence, the individual seems best regarded as a pawn in the game, to be sacrificed when the greater interest of the species as a whole requires it. To put it in a slightly more respectable way, a group, such as a species or a population within a species, whose individual members are prepared to sacrifice themselves for the welfare of the group, may be less likely to go extinct than a rival group whose individual members place their own selfish interests first. Therefore the world becomes populated mainly by groups consisting of self-sacrificing individuals. This is the theory of 'group selection', long assumed to be true by biologists not familiar with the details of evolutionary theory, brought out into the open in a famous book by V. C. Wynne-Edwards, and popularized by Robert Ardrey in *The Social Contract*. The orthodox alternative is normally called 'individual selection', although I personally prefer to speak of gene selection.

The quick answer of the 'individual selectionist' to the argument just put might go something like this. Even in the group of altruists, there will almost certainly be a dissenting minority who refuse to make any sacrifice. If there is just one selfish rebel, prepared to exploit the altruism of the rest, then he, by definition, is more likely than they are to survive and have children. Each of these children will tend to inherit his selfish traits. After several generations of this natural selection, the 'altruistic group' will be over-run by selfish individuals, and will be indistinguishable from the selfish group. Even if we grant the improbable chance existence initially of pure altruistic groups without any rebels, it is very difficult to see what is to stop selfish individuals migrating in from neighbouring selfish groups, and, by inter-marriage, contaminating the purity of the altruistic groups.

The individual-selectionist would admit that groups do indeed die out, and that whether or not a group goes extinct may be influenced by the behaviour of the individuals in that group. He might even admit that *if only* the individuals in a group had the gift of foresight they could see that in the long run their own best interests lay in restraining their selfish greed, to prevent the destruction of the whole group. How many times must this have been said in recent years to the working people of Britain? But group extinction is a slow process compared with the rapid cut and thrust of individual competition. Even while the group is going slowly and inexorably downhill, selfish individuals prosper in the short term at the expense of altruists. The citizens of Britain may or may not be blessed with foresight, but evolution is blind to the future.

Although the group-selection theory now commands little support within the ranks of those professional biologists who understand evolution, it does have great intuitive appeal. Successive generations of zoology students are surprised, when they come up from school, to find that it is not the orthodox point of view. For this they are hardly to be blamed, for in the *Nuffield Biology Teachers' Guide*, written for advanced level biology schoolteachers in Britain, we find the following: 'In higher animals, behaviour may take the form of individual suicide to ensure the survival of the species.' The anonymous author of this guide is blissfully ignorant of the fact that he has said something controversial. In this respect he is in Nobel Prize-winning company. Konrad Lorenz, in *On Aggression*, speaks of the 'species preserving' functions of aggressive behaviour, one of these functions being to make sure that only the fittest individuals are allowed to breed. This is a gem of a circular argument, but the point I am making here is that the group selection idea is so deeply ingrained that Lorenz, like the author of the *Nuffield Guide*, evidently did not realize that his statements contravened orthodox Darwinian theory.

I recently heard a delightful example of the same thing on an otherwise excellent B.B.C. television programme about Australian spiders. The 'expert' on the programme observed that the vast majority of baby spiders end up as prey for other species, and she then went on to say: 'Perhaps this is the real purpose of their existence, as only a few need to survive in order for the species to be preserved'!

Robert Ardrey, in *The Social Contract*, used the group-selection theory to account for the whole of social order in general. He clearly sees man as a species which has strayed from the path of animal righteousness. Ardrey at least did his homework. His decision to disagree with orthodox theory was a conscious one, and for this he deserves credit.

Perhaps one reason for the great appeal of the group-selection theory is that it is thoroughly in tune with the moral and political ideals which most of us share. We may frequently behave selfishly as individuals, but in our more idealistic moments we honour and admire those who put the welfare of others first. We get a bit muddled over how widely we want to interpret the word 'others', though. Often altruism within a group goes with selfishness between groups. This is a basis of trade unionism. At another level the nation is a major beneficiary of our altruistic self-sacrifice, and young men are expected to die as individuals for the greater glory of their country as a whole. Moreover, they are encouraged to kill other individuals about whom nothing is known except that they belong to a different nation. (Curiously, peace-time appeals for individuals to make some small sacrifice in the rate at which they increase their standard of living seem to be less effective than war-time appeals for individuals to lay down their lives.)

Recently there has been a reaction against racialism and patriotism, and a tendency to substitute the whole human species as the object of our fellow feeling. This humanist broadening of the target of our altruism has an interesting corollary, which again seems to buttress the 'good of the species' idea in evolution. The politically liberal, who are normally the most convinced spokesmen of the species ethic, now often have the greatest scorn for those who have gone a little further in widening their altruism, so that it includes other species. If I say that I am more interested in preventing the slaughter of large whales than I am in improving housing conditions for people, I am likely to shock some of my friends.

The feeling that members of one's own species deserve special moral consideration as compared with members of other species is old and deep. Killing people outside war is the most seriously-regarded crime ordinarily committed. The only thing more strongly forbidden by our culture is eating people (even if they are already dead). We enjoy eating members of other species, however. Many of us shrink from judicial execution of even the most horrible human criminals, while we cheerfully countenance the shooting without trial of fairly mild animal pests. Indeed we kill members of other harmless species as a means of recreation and amusement. A human foetus, with no more human feeling than an amoeba, enjoys a reverence and legal protection far in excess of those granted to an adult chimpanzee. Yet the chimp feels and thinks and—according to recent experimental evidence —may even be capable of learning a form of human language. The foetus belongs to our own species, and is instantly accorded special privileges and rights because of it. Whether the ethic of 'speciesism', to use Richard Ryder's term, can be put on a logical footing any more sound than that of 'racism', I do not know. What I do know is that it has no proper basis in evolutionary biology.

The muddle in human ethics over the level at which altruism is desirable—family, nation, race, species, or all living things—is mirrored by a parallel muddle in biology over the level at which altruism is to be expected according to the theory of evolution. Even the group-selectionist would not be surprised to find members of rival groups being nasty to each other: in this way, like trade unionists or soldiers, they are favouring their own group in the struggle for limited resources. But then it is worth asking how the group-selectionist decides *which* level is the important one. If selection goes on between groups within a species, and between species, why should it not also go on between larger groupings? Species are grouped together into genera, genera into orders, and orders into classes. Lions and antelopes are both members of the class Mammalia, as are we. Should we then not expect lions to refrain from killing antelopes, 'for the good of the mammals'? Surely they should hunt birds or reptiles instead, in order to prevent the extinction of the class. But then, what of the need to perpetuate the whole phylum of vertebrates?

It is all very well for me to argue by *reductio ad absurdum*, and to point to the difficulties of the group-selection theory, but the apparent existence of individual altruism still has to be explained. Ardrey goes so far as to say that group selection is the only possible explanation for behaviour such as 'stotting' in Thomson's gazelles. This vigorous and conspicuous leaping in front of a predator is analogous to bird alarm calls, in that it seems to warn companions of danger while apparently calling the predator's attention to the stotter himself. We have a responsibility to explain stotting Tommies and all similar phenomena, and this is something I am going to face in later chapters.

Before that I must argue for my belief that the best way to look at evolution is in terms of selection occurring at the lowest level of all. In this belief I am heavily influenced by G. C. Williams's great book *Adaptation and Natural Selection*. The central idea I shall make use of was foreshadowed by A. Weismann in pre-gene days at the turn of the century—his doctrine of the 'continuity of the germ-plasm'. I shall argue that the fundamental unit of selection, and therefore of self-interest, is not the species, nor the group, nor even, strictly, the individual. It is the gene, the unit of heredity. To some biologists this may sound at first like an extreme view. I hope when they see in what sense I mean it they will agree that it is, in substance, orthodox, even if it is expressed in an unfamiliar way. The argument takes time to develop, and we must begin at the beginning, with the very origin of life itself.

2. The replicators

In the beginning was simplicity. It is difficult enough explaining how even a simple universe began. I take it as agreed that it would be even harder to explain the sudden springing up, fully armed, of complex order—life, or a being capable of creating life. Darwin's theory of evolution by natural selection is satisfying because it shows us a way in which simplicity could change into complexity, how unordered atoms could group themselves into ever more complex patterns until they ended up manufacturing people. Darwin provides a solution, the only feasible one so far suggested, to the deep problem of our existence. I will try to explain the great theory in a more general way than is customary, beginning with the time before evolution itself began.

Darwin's 'survival of the fittest' is really a special case of a more general law of *survival of the stable*. The universe is populated by stable things. A stable thing is a collection of atoms which is permanent enough or common enough to deserve a name. It may be a unique collection of atoms, such as the Matterhorn, which lasts long enough to be worth naming. Or it may be a *class* of entities, such as rain drops, which come into existence at a sufficiently high rate to deserve a collective name, even if any one of them is short-lived. The things which we see around us, and which we think of as needing explanation—rocks, galaxies, ocean waves—are all, to a greater or lesser extent, stable patterns of atoms. Soap bubbles tend to be spherical because this is a stable configuration for thin films filled with gas. In a space-craft, water is also stable in spherical globules, but on earth, where there is gravity, the stable surface for standing water is flat and horizontal. Salt crystals tend to be cubes because this is a stable way of packing sodium and chloride ions together. In the sun the simplest atoms of all, hydrogen atoms, are fusing to form helium atoms, because in the conditions which prevail there the helium configuration is more stable. Other even more complex atoms are being formed in stars all over the universe, and were formed in the 'big bang' which, according to the prevailing theory, initiated the universe. This is originally where the elements on our world came from.

Sometimes when atoms meet they link up together in chemical reaction to form molecules, which may be more or less stable. Such molecules can be very large. A crystal such as a diamond can be regarded as a single molecule, a proverbially stable one in this case, but also a very simple one since its internal atomic structure is endlessly repeated. In modern living organisms there are other large molecules which are highly complex, and their complexity shows itself on several levels. The haemoglobin of our blood is a typical protein molecule. It is built up from chains of smaller molecules, amino acids, each containing a few dozen atoms arranged in a precise pattern. In the haemoglobin molecule there are 574 amino acid molecules. These are arranged in four chains, which twist around each other to form a globular three-dimensional structure of bewildering complexity. A model of a haemoglobin molecule looks rather like a dense thornbush. But unlike a real thornbush it is not a haphazard approximate pattern but a definite invariant structure, identically repeated, with not a twig nor a twist out of place, over six thousand million million million times in an average human body. The precise thornbush shape of a protein molecule such as haemoglobin is stable in the sense that two chains consisting of the same sequences of amino acids will tend, like two springs, to come to rest in exactly the same three-dimensional coiled pattern. Haemoglobin thornbushes are springing into their 'preferred' shape in your body at a rate of about four hundred million million per second, and others are being destroyed at the same rate.

Haemoglobin is a modern molecule, used to illustrate the principle that atoms tend to fall into stable patterns. The point that is relevant here is that, before the coming of life on earth, some rudimentary evolution of molecules could have occurred by ordinary processes of physics and chemistry. There is no need to think of design or purpose or directedness. If a group of atoms in the presence of energy falls into a stable pattern it will tend to stay that way. The earliest form of natural selection was simply a selection of stable forms and a rejection of unstable ones. There is no mystery about this. It had to happen by definition.

From this, of course, it does not follow that you can explain the existence of entities as complex as man by exactly the same principles on their own. It is no good taking the right number of atoms and shaking them together with some external energy till they happen to fall into the right pattern, and out drops Adam! You may make a molecule consisting of a few dozen atoms like that, but a man consists of over a thousand million million million million atoms. To try to make a man, you would have to work at your biochemical cocktail-shaker for a period so long that the entire age of the universe would seem like an eye-blink, and even then you would not succeed. This is where Darwin's theory, in its most general form, comes to the rescue. Darwin's theory takes over from where the story of the slow building up of molecules leaves off.

The account of the origin of life which I shall give is necessarily speculative; by definition, nobody was around to see what happened. There are a number of rival theories, but they all have certain features in common. The simplified account I shall give is probably not too far from the truth.

We do not know what chemical raw materials were abundant on earth before the coming of life, but among the plausible possibilities are water, carbon dioxide, methane, and ammonia: all simple compounds known to be present on at least some of the other planets in our solar system. Chemists have tried to imitate the chemical conditions of the young earth. They have put these simple substances in a flask and supplied a source of energy such as ultraviolet light or electric sparks—artificial simulation of primordial lightning. After a few weeks of this, something interesting is usually found inside the flask: a weak brown soup containing a large number of molecules more complex than the ones originally put in. In particular, amino acids have been found—the building blocks of proteins, one of the two great classes of biological molecules. Before these experiments were done, naturally-occurring amino acids would have been thought of as diagnostic of the presence of life. If they had been detected on, say Mars, life on that planet would have seemed a near certainty. Now, however, their existence need imply only the presence of a few simple gases in the atmosphere and some volcanoes, sunlight, or thundery weather. More recently, laboratory simulations of the chemical conditions of earth before the coming of life have yielded organic substances called purines and pyrimidines. These are building blocks of the genetic molecule, DNA itself.

Processes analogous to these must have given rise to the 'primeval soup' which biologists and chemists believe constituted the seas some three to four thousand million years ago. The organic substances became locally concentrated, perhaps in drying scum round the shores, or in tiny suspended droplets. Under the further influence of energy such as ultraviolet light from the sun, they combined into larger molecules. Nowadays large organic molecules would not last long enough to be noticed: they would be quickly absorbed and broken down by bacteria or other living creatures. But bacteria and the rest of us are late-comers, and in those days large organic molecules could drift unmolested through the thickening broth.

At some point a particularly remarkable molecule was formed by accident. We will call it the *Replicator*. It may not necessarily have been the biggest or the most complex molecule around, but it had the extraordinary property of being able to create copies of itself. This may seem a very unlikely sort of accident to happen. So it was. It was exceedingly improbable. In the lifetime of a man, things which are that improbable can be treated for practical purposes as impossible. That is why you will never win a big prize on the football pools. But in our human estimates of what is probable and what is not, we are not used to dealing in hundreds of millions of years. If you filled in pools coupons every week for a hundred million years you would very likely win several jackpots.

Actually a molecule which makes copies of itself is not as difficult to imagine as it seems at first, and it only had to arise once. Think of the replicator as a mould or template. Imagine it as a large molecule consisting of a complex chain of various sorts of building block molecules. The small building blocks were abundantly available in the soup surrounding the replicator. Now suppose that each building block has an affinity for its own kind. Then whenever a building block from out in the soup lands up next to a part of the replicator for which it has an affinity, it will tend to stick there. The building blocks which attach themselves in this way will automatically be arranged in a sequence which mimics that of the replicator itself. It is easy then to think of them joining up to form a stable chain just as in the formation of the original replicator. This process could continue as a progressive stacking up, layer upon layer. This is how crystals are formed. On the other hand, the two chains might split apart, in which case we have two replicators, each of which can go on to make further copies.

A more complex possibility is that each building block has affinity not for its own kind, but reciprocally for one particular other kind. Then the replicator would act as a template not for an identical copy, but for a kind of 'negative', which would in its turn re-make an exact copy of the original positive. For our purposes it does not matter whether the original replication process was positive–negative or positive–positive, though it is worth remarking that the modern equivalents of the first replicator, the DNA molecules, use positive–negative replication. What does matter is that suddenly a new kind of 'stability' came into the world. Previously it is probable that no particular kind of complex molecule was very abundant in the soup, because each was dependent on building blocks happening to fall by luck into a particular stable configuration. As soon as the replicator was born it must have spread its copies rapidly throughout the seas, until the smaller building block molecules became a scarce resource, and other larger molecules were formed more and more rarely.

So we seem to arrive at a large population of identical replicas. But now we must mention an important property of any copying process: it is not perfect. Mistakes will happen. I hope there are no misprints in this book, but if you look carefully you may find one or two. They will probably not seriously distort the meaning of the sentences, because they will be 'first generation' errors. But imagine the days before printing, when books such as the Gospels were copied by hand. All scribes, however careful, are bound to make a few errors, and some are not above a little wilful 'improvement'. If they all copied from a single master original, meaning would not be greatly perverted. But let copies be made from other copies, which in their turn were made from other copies, and errors will start to become cumulative and serious. We tend to regard erratic copying as a bad thing, and in the case of human documents it is hard to think of examples where errors can be described as improvements. I suppose the scholars of the Septuagint could at least be said to have started something big when they mistranslated the Hebrew word for 'young woman' into the Greek word for 'virgin', coming up with the prophecy: 'Behold a virgin shall conceive and bear a son. . . .' Anyway, as we shall see, erratic copying in biological replicators can in a real sense give rise to improvement, and it was essential for the progressive evolution of life that some errors were made. We do not know how accurately the original replicator molecules made their copies. Their modern descendants, the DNA molecules, are astonishingly faithful compared with the most high-fidelity human copying process, but even they occasionally make mistakes, and it is ultimately these mistakes which make evolution possible. Probably the original replicators were far more erratic, but in any case we may be sure that mistakes were made, and these mistakes were cumulative.

As mis-copyings were made and propagated, the primeval soup became filled by a population not of identical replicas, but of several varieties of replicating molecules, all 'descended' from the same ancestor. Would some varieties have been more numerous than others? Almost certainly yes. Some varieties would have been inherently more stable than others. Certain molecules, once formed, would be less likely than others to break up again. These types would become relatively numerous in the soup, not only as a direct logical consequence of their 'longevity', but also because they would have a long time available for making copies of themselves. Replicators of high longevity would therefore tend to become more numerous and, other things being equal, there would have been an 'evolutionary trend' towards greater longevity in the population of molecules.

But other things were probably not equal, and another property of a replicator variety which must have had even more importance in spreading it through the population was speed of replication or 'fecundity'. If replicator molecules of type A make copies of themselves on average once a week while those of type B make copies of themselves once an hour, it is not difficult to see that pretty soon type A molecules are going to be far outnumbered, even if they 'live' much longer than B molecules. There would therefore probably have been an 'evolutionary trend' towards higher 'fecundity' of molecules in the soup. A third characteristic of replicator molecules which would have been positively selected is accuracy of replication. If molecules of type X and type Y last the same length of time and replicate at the same rate, but X makes a mistake on average every tenth replication while Y makes a mistake only every hundredth replication, Y will obviously become more numerous. The X contingent in the population loses not only the errant 'children' themselves, but also all their descendants, actual or potential.

If you already know something about evolution, you may find something slightly paradoxical about the last point. Can we reconcile the idea that copying errors are an essential prerequisite for evolution to occur, with the statement that natural selection favours high copying-fidelity? The answer is that although evolution may seem, in some vague sense, a 'good thing', especially since we are the product of it, nothing actually 'wants' to evolve. Evolution is something that happens, willy-nilly, in spite of all the efforts of the replicators (and nowadays of the genes) to prevent it happening. Jacques Monod made this point very well in his Herbert Spencer lecture, after wryly remarking: 'Another curious aspect of the theory of evolution is that everybody thinks he understands it!'

To return to the primeval soup, it must have become populated by stable varieties of molecule; stable in that either the individual molecules lasted a long time, or they replicated rapidly, or they replicated accurately. Evolutionary trends toward these three kinds of stability took place in the following sense: if you had sampled the soup at two different times, the later sample would have contained a higher proportion of varieties with high longevity/fecundity/copying-fidelity. This is essentially what a biologist means by evolution when he is speaking of living creatures, and the mechanism is the same—natural selection.

Should we then call the original replicator molecules 'living'? Who cares? I might say to you 'Darwin was the greatest man who has ever lived', and you might say 'No, Newton was', but I hope we would not prolong the argument. The point is that no conclusion of substance would be affected whichever way our argument was resolved. The facts of the lives· and achievements of Newton and Darwin remain totally unchanged whether we label them 'great' or not. Similarly, the story of the replicator molecules probably happened something like the way I am telling it, regardless of whether we choose to call them 'living'. Human suffering has been caused because too many of us cannot grasp that words are only tools for our use, and that the mere presence in the dictionary of a word like 'living' does not mean it necessarily has to refer to something definite in the real world. Whether we call the early replicators living or not, they were the ancestors of life; they were our founding fathers.

The next important link in the argument, one which Darwin himself laid stress on (although he was talking about animals and plants, not molecules) is *competition*. The primeval soup was not capable of supporting an infinite number of replicator molecules. For one thing, the earth's size is finite, but other limiting factors must also have been important. In our picture of the replicator acting as a template or mould, we supposed it to be bathed in a soup rich in the small building block molecules necessary to make copies. But when the replicators became numerous, building blocks must have been used up at such a rate that they became a scarce and precious resource. Different varieties or strains of replicator must have competed for them. We have considered the factors which would have increased the numbers of favoured kinds of replicator. We can now see that less-favoured varieties must actually have become *less* numerous because of competition, and ultimately many of their lines must have gone extinct. There was a struggle for existence among replicator varieties. They did not know they were struggling, or worry about it; the struggle was conducted without any hard feelings, indeed without feelings of any kind. But they were struggling, in the sense that any mis-copying which resulted in a new higher level of stability, or a new way of reducing the stability of rivals, was automatically preserved and multiplied. The process of improvement was cumulative. Ways of increasing stability and of decreasing rivals' stability became more elaborate and more efficient. Some of them may even have 'discovered' how to break up molecules of rival varieties chemically, and to use the building blocks so released for making their own copies. These proto-carnivores simultaneously obtained food and removed competing rivals. Other replicators perhaps discovered how to protect themselves, either chemically, or by building a physical wall of protein around themselves. This may have been how the first living cells appeared. Replicators began not merely to exist, but to construct for themselves containers, vehicles for their continued existence. The replicators which survived were the ones which built *survival machines* for themselves to live in. The first survival machines probably consisted of nothing more than a protective coat. But making a living got steadily harder as new rivals arose with better and more effective survival machines. Survival machines got bigger and more elaborate, and the process was cumulative and progressive.

Was there to be any end to the gradual improvement in the techniques and artifices used by the replicators to ensure their own continuance in the world? There would be plenty of time for improvement. What weird engines of self-preservation would the millennia bring forth? Four thousand million years on, what was to be the fate of the ancient replicators? They did not die out, for they are past masters of the survival arts. But do not look for them floating loose in the sea; they gave up that cavalier freedom long ago. Now they swarm in huge colonies, safe inside gigantic lumbering robots, sealed off from the outside world, communicating with it by tortuous indirect routes, manipulating it by remote control. They are in you and in me; they created us, body and mind; and their preservation is the ultimate rationale for our existence. They have come a long way, those replicators. Now they go by the name of genes, and we are their survival machines.

3. Immortal coils

WE are survival machines, but 'we' does not mean just people. It embraces all animals, plants, bacteria, and viruses. The total number of survival machines on earth is very difficult to count and even the total number of species is unknown. Taking just insects alone, the number of living species has been estimated at around three million, and the number of individual insects may be a million million million.

Different sorts of survival machine appear very varied on the outside and in their internal organs. An octopus is nothing like a mouse, and both are quite different from an oak tree. Yet in their fundamental chemistry they are rather uniform, and, in particular, the replicators which they bear, the genes, are basically the same kind of molecule in all of us—from bacteria to elephants. We are all survival machines for the same kind of replicator—molecules called DNA—but there are many different ways of making a living in the world, and the replicators have built a vast range of machines to exploit them. A monkey is a machine which preserves genes up trees, a fish is a machine which preserves genes in the water; there is even a small worm which preserves genes in German beer mats. DNA works in mysterious ways.

For simplicity I have given the impression that modern genes, made of DNA, are much the same as the first replicators in the primeval soup. It does not matter for the argument, but this may not really be true. The original replicators may have been a related kind of molecule to DNA, or they may have been totally different. In the latter case we might say that their survival machines must have been seized at a later stage by DNA. If so, the original replicators were utterly destroyed, for no trace of them remains in modern survival machines. Along these lines, A. G. Cairns-Smith has made the intriguing suggestion that our ancestors, the first replicators, may have been not organic molecules at all, but inorganic crystals—minerals, little bits of clay. Usurper or not, DNA is in undisputed charge today, unless, as I tentatively suggest in the final chapter, a new seizure of power is now just beginning.

A DNA molecule is a long chain of building blocks, small molecules called nucleotides. Just as protein molecules are chains of amino acids, so DNA molecules are chains of nucleotides. A DNA molecule is too small to be seen, but its exact shape has been ingeniously worked out by indirect means. It consists of a pair of nucleotide chains twisted together in an elegant spiral; the 'double helix'; the 'immortal coil'. The nucleotide building blocks come in only four different kinds, whose names may be shortened to *A*, *T*, *C*, and *G*. These are the same in all animals and plants. What differs is the order in which they are strung together. A *G* building block from a man is identical in every particular to a *G* building block from a snail. But the *sequence* of building blocks in a man is not only different from that in a snail. It is also different—though less so—from the sequence in every other man (except in the special case of identical twins).

Our DNA lives inside our bodies. It is not concentrated in a particular part of the body, but is distributed among the cells. There are about a thousand million million cells making up an average human body, and, with some exceptions which we can ignore, every one of those cells contains a complete copy of that body's DNA. This DNA can be regarded as a set of instructions for how to make a body, written in the *A*, *T*, *C*, *G* alphabet of the nucleotides. It is as though, in every room of a gigantic building, there was a book-case containing the architect's plans for the entire building. The 'book-case' in a cell is called the nucleus. The architect's plans run to 46 volumes in man—the number is different in other species. The 'volumes' are called chromosomes. They are visible under a microscope as long threads, and the genes are strung out along them in order. It is not easy, indeed it may not even be meaningful, to decide where one gene ends and the next one begins. Fortunately, as this chapter will show, this does not matter for our purposes.

I shall make use of the metaphor of the architect's plans, freely mixing the language of the metaphor with the language of the real thing. 'Volume' will be used interchangeably with chromosome.

'Page' will provisionally be used interchangeably with gene, although the division between genes is less clear-cut than the division between the pages of a book. This metaphor will take us quite a long way. When it finally breaks down I shall introduce other metaphors. Incidentally, there is of course no 'architect'. The DNA instructions have been assembled by natural selection.

DNA molecules do two important things. Firstly they replicate, that is to say they make copies of themselves. This has gone on non-stop ever since the beginning of life, and the DNA molecules are now very good at it indeed. As an adult, you consist of a thousand million million cells, but when you were first conceived you were just a single cell, endowed with one master copy of the architect's plans. This cell divided into two, and each of the two cells received its own copy of the plans. Successive divisions took the number of cells up to 4, 8, 16, 32, and so on into the billions. At every division the DNA plans were faithfully copied, with scarcely any mistakes.

It is one thing to speak of the duplication of DNA. But if the DNA is really a set of plans for building a body, how are the plans put into practice? How are they translated into the fabric of the body? This brings me to the second important thing DNA does. It indirectly supervises the manufacture of a different kind of molecule—protein. The haemoglobin which was mentioned in the last chapter is just one example of the enormous range of protein molecules. The coded message of the DNA, written in the four-letter nucleotide alphabet, is translated in a simple mechanical way into another alphabet. This is the alphabet of amino acids which spells out protein molecules.

Making proteins may seem a far cry from making a body, but it is the first small step in that direction. Proteins not only constitute much of the physical fabric of the body; they also exert sensitive control over all the chemical processes inside the cell, selectively turning them on and off at precise times and in precise places. Exactly how this eventually leads to the development of a baby is a story which it will take decades, perhaps centuries, for embryologists to work out. But it is a fact that it does. Genes do indirectly control the manufacture of bodies, and the influence is strictly one way: acquired characteristics are not inherited. No matter how much knowledge and wisdom you acquire during your life, not one jot will be passed on to your children by genetic means. Each new generation starts from scratch. A body is the genes' way of preserving the genes unaltered.

The evolutionary importance of the fact that genes control embryonic development is this: it means that genes are at least partly responsible for their own survival in the future, because their survival depends on the efficiency of the bodies in which they live and which they helped to build. Once upon a time, natural selection consisted of the differential survival of replicators floating free in the primeval soup. Now, natural selection favours replicators which are good at building survival machines, genes which are skilled in the art of controlling embryonic development. In this, the replicators are no more conscious or purposeful than they ever were. The same old processes of automatic selection between rival molecules by reason of their longevity, fecundity, and copying-fidelity, still go on as blindly and as inevitably as they did in the far-off days. Genes have no foresight. They do not plan ahead. Genes just *are*, some genes more so than others, and that is all there is to it. But the qualities which determine a gene's longevity and fecundity are not so simple as they were. Not by a long way.

In recent years—the last six hundred million or so—the replicators have achieved notable triumphs of survival-machine technology such as the muscle, the heart, and the eye (evolved several times independently). Before that, they radically altered fundamental features of their way of life as replicators, which must be understood if we are to proceed with the argument.

The first thing to grasp about a modern replicator is that it is highly gregarious. A survival machine is a vehicle containing not just one gene but many thousands. The manufacture of a body is a cooperative venture of such intricacy that it is almost impossible to disentangle the contribution of one gene from that of another. A given gene will have many different effects on quite different parts of the body. A given part of the body will be influenced by many genes, and the effect of any one gene depends on interaction with many others. Some genes act as master genes controlling the operation of a cluster of other genes. In terms of the analogy, any given page of the plans makes reference to many different parts of the building; and each page makes sense only in terms of cross-references to numerous other pages.

This intricate inter-dependence of genes may make you wonder why we use the word 'gene' at all. Why not use a collective noun like 'gene complex'? The answer is that for many purposes that is indeed quite a good idea. But if we look at things in another way, it does make sense too to think of the gene complex as being divided up into discrete replicators or genes. This arises because of the phenomenon of sex. Sexual reproduction has the effect of mixing and shuffling genes. This means that any one individual body is just a temporary vehicle for a short-lived combination of genes. The *combination* of genes which is any one individual may be short-lived, but the genes themselves are potentially very long-lived. Their paths constantly cross and re-cross down the generations. One gene may be regarded as a unit which survives through a large number of successive individual bodies. This is the central argument which will be developed in this chapter. It is an argument which some of my most respected colleagues obstinately refuse to agree with, so you must forgive me if I seem to labour it! First I must briefly explain the facts of sex.

I said that the plans for building a human body are spelt out in 46 volumes. In fact this was an over-simplification. The truth is rather bizarre. The 46 chromosomes consist of 23 *pairs* of chromosomes. We might say that, filed away in the nucleus of every cell, are two alternative sets of 23 volumes of plans. Call them Volume 1a and Volume 1b, Volume 2a and Volume 2b etc., down to Volume 23a and Volume 23b. Of course the identifying numbers I use for volumes and, later, pages, are purely arbitrary.

We receive each chromosome intact from one of our two parents, in whose testis or ovary it was assembled. Volumes 1a, 2a, 3a, . . . came, say, from the father. Volumes 1b, 2b, 3b, . . . came from the mother. It is very difficult in practice, but in theory you could look with a microscope at the 46 chromosomes in any one of your cells, and pick out the 23 that came from your father and the 23 that came from your mother.

The paired chromosomes do not spend all their lives physically in contact with each other, or even near each other. In what sense then are they 'paired'? In the sense that each volume coming originally from the father can be regarded, page for page, as a direct alternative to one particular volume coming originally from the mother. For instance, Page 6 of Volume 13a and Page 6 of Volume 13b might both be 'about' eye colour; perhaps one says 'blue' while the other says 'brown'.

Sometimes the two alternative pages are identical, but in other cases, as in our example of eye colour, they differ. If they make contradictory 'recommendations', what does the body do? The answer varies. Sometimes one reading prevails over the other. In the eye colour example just given, the person would actually have brown eyes: the instructions for making blue eyes would be ignored in the building of the body, though this does not stop them being passed on to future generations. A gene which is ignored in this way is called *recessive*. The opposite of a recessive gene is a *dominant* gene. The gene for brown eyes is dominant to the gene for blue eyes. A person has blue eyes only if both copies of the relevant page are unanimous in recommending blue eyes. More usually when two alternative genes are not identical, the result is some kind of compromise—the body is built to an inter-mediate design or something completely different.

When two genes, like the brown eye and the blue eye gene, are rivals for the same slot on a chromosome, they are called *alleles* of each other. For our purposes, the word allele is synonymous with rival. Imagine the volumes of architect's plans as being loose-leaf binders, whose pages can be detached and interchanged. Every Volume 13 must have a Page 6, but there are several possible Page 6s which could go in the binder between Page 5 and Page 7. One version says 'blue eyes', another possible version says 'brown eyes'; there may be yet other versions in the population at large which spell out other colours like green. Perhaps there are half a dozen alternative alleles sitting in the Page 6 position on the 13th chromosomes scattered around the population as a whole. Any given person only has two Volume 13 chromosomes. Therefore he can have a maximum of two alleles in the Page 6 slot. He may, like a blue-eyed person, have two copies of the same allele, or he may have any two alleles chosen from the half dozen alternatives available in the population at large.

You cannot, of course, literally go and choose your genes from a pool of genes available to the whole population. At any given time all the genes are tied up inside individual survival machines.

Our genes are doled out to us at conception, and there is nothing we can do about this. Nevertheless, there is a sense in which, in the long term, the genes of the population in general can be regarded as a *gene pool*. This phrase is in fact a technical term used by geneticists. The gene pool is a worthwhile abstraction because sex mixes genes up, albeit in a carefully organized way. In particular, something like the detaching and interchanging of pages and wads of pages from loose-leaf binders really does go on, as we shall presently see.

I have described the normal division of a cell into two new cells, each one receiving a complete copy of all 46 chromosomes. This normal cell division is called *mitosis*. But there is another kind of cell division called *meiosis*. This occurs only in the production of the sex cells; the sperms or eggs. Sperms and eggs are unique among our cells in that, instead of containing 46 chromosomes, they contain only 23. This is, of course, exactly half of 46—convenient when they fuse in sexual fertilization to make a new individual! Meiosis is a special kind of cell division, taking place only in testicles and ovaries, in which a cell with the full double set of 46 chromosomes divides to form sex cells with the single set of 23 (all the time using the human numbers for illustration).

A sperm, with its 23 chromosomes, is made by the meiotic division of one of the ordinary 46-chromosome cells in the testicle. Which 23 are put into any given sperm cell? It is clearly important that a sperm should not get just any old 23 chromosomes: it mustn't end up with two copies of Volume 13 and none of Volume 17. It would theoretically be possible for an individual to endow one of his sperms with chromosomes which came, say, entirely from his mother; that is Volume 1b, 2b, 3b, . . . , 23b. In this unlikely event, a child conceived by the sperm would inherit half her genes from her paternal grandmother, and none from her paternal grandfather. But in fact this kind of gross, whole-chromosome distribution does not happen. The truth is rather more complex. Remember that the volumes (chromosomes) are to be thought of as loose-leaf binders. What happens is that, during the manufacture of the sperm, single pages, or rather multi-page chunks, are detached and swapped with the corresponding chunks from the alternative volume. So, one particular sperm cell might make up its Volume 1 by taking the first 65 pages from Volume 1a, and pages 66 to the end from Volume 1b. This sperm cell's other 22 volumes would be made up in a similar way. Therefore every sperm cell made by an individual is unique, even though all his sperms assembled their 23 chromosomes from bits of the same set of 46 chromosomes. Eggs are made in a similar way in ovaries, and they too are all unique.

The real-life mechanics of this mixing are fairly well understood. During the manufacture of a sperm (or egg), bits of each paternal chromosome physically detach themselves and change places with exactly corresponding bits of maternal chromosome. (Remember that we are talking about chromosomes which came originally from the parents of the individual making the sperm, i.e., from the paternal grandparents of the child who is eventually conceived by the sperm). The process of swapping bits of chromosome is called *crossing over*. It is very important for the whole argument of this book. It means that if you got out your microscope and looked at the chromosomes in one of your own sperms (or eggs if you are female) it would be a waste of time trying to identify chromosomes which originally came from your father and chromosomes which originally came from your mother. (This is in marked contrast to the case of ordinary body cells (see page 26).) Any one chromosome in a sperm would be a patchwork, a mosaic of maternal genes and paternal genes.

The metaphor of the page for the gene starts to break down here. In a loose-leaf binder a whole page may be inserted, removed or exchanged, but not a fraction of a page. But the gene complex is just a long string of nucleotide letters, not divided into discrete pages in an obvious way at all. To be sure, there are special symbols for END OF PROTEIN CHAIN MESSAGE and START OF PROTEIN CHAIN MESSAGE written in the same four-letter alphabet as the protein messages themselves. In between these two punctuation marks are the coded instructions for making one protein. If we wish, we can define a single gene as a sequence of nucleotide letters lying between a START and an END symbol, and coding for one protein chain. The word *cistron* has been used for a unit defined in this way, and some people use the word gene interchangeably with cistron. But crossing-over does not respect boundaries between cistrons. Splits may occur within cistrons as well as between them. It is as though the architect's plans were written out, not on discrete pages, but on 46 rolls of ticker tape. Cistrons are not of fixed length. The only way to tell where one cistron ends and the next begins would be to read the symbols on the tape, looking for END OF MESSAGE and START OF MESSAGE symbols. Crossing-over is represented by taking matching paternal and maternal tapes, and cutting and exchanging matching portions, regardless of what is written on them.

In the title of this book the word gene means not a single cistron but something more subtle. My definition will not be to everyone's taste, but there is no universally agreed definition of a gene. Even if there were, there is nothing sacred about definitions. We can define a word how we like for our own purposes, provided we do so clearly and unambiguously. The definition I want to use comes from G. C. Williams. A gene is defined as any portion of chromosomal material which potentially lasts for enough generations to serve as a unit of natural selection. In the words of the previous chapter, a gene is a replicator with high copying-fidelity. Copying-fidelity is another way of saying longevity-in-the-form-of-copies and I shall abbreviate this simply to longevity. The definition will take some justifying.

On any definition, a gene has to be a portion of a chromosome. The question is, how big a portion—how much of the ticker tape? Imagine any sequence of adjacent code-letters on the tape. Call the sequence a *genetic unit*. It might be a sequence of only ten letters within one cistron; it might be a sequence of eight cistrons; it might start and end in mid-cistron. It will overlap with other genetic units. It will include smaller units, and it will form part of larger units. No matter how long or short it is, for the purposes of the present argument, this is what we are calling a genetic unit. It is just a length of chromosome, not physically differentiated from the rest of the chromosome in any way.

Now comes the important point. The shorter a genetic unit is, the longer—in generations—it is likely to live. In particular, the less likely it is to be split by any one crossing-over. Suppose a whole chromosome is, on average, likely to undergo one crossover every time a sperm or egg is made by meiotic division, and this cross-over can happen anywhere along its length. If we consider a very large genetic unit, say half the length of the chromosome, there is a 50 per cent chance that the unit will be split at each meiosis. If the genetic unit we are considering is only 1 per cent of the length of the chromosome, we can assume that it has only a 1 per cent chance of being split in any one meiotic division. This means that the unit can expect to survive for a large number of generations in the individual's descendants. A single cistron is likely to be much less than 1 per cent of the length of a chromosome. Even a group of several neighbouring cistrons can expect to live many generations before being broken up by crossing over.

The average life-expectancy of a genetic unit can conveniently be expressed in generations, which can in turn be translated into years. If we take a whole chromosome as our presumptive genetic unit, its life story lasts for only one generation. Suppose it is your chromosome number 8a, inherited from your father. It was created inside one of your father's testicles, shortly before you were conceived. It had never existed before in the whole history of the world. It was created by the meiotic shuffling process, forged by the coming together of pieces of chromosome from your paternal grandmother and your paternal grandfather. It was placed inside one particular sperm, and it was unique. The sperm was one of several millions, a vast armada of tiny vessels, and together they sailed into your mother. This particular sperm (unless you are a non-identical twin) was the only one of the flotilla which found harbour in one of your mother's eggs—that is why you exist. The genetic unit we are considering, your chromosome number 8a, set about replicating itself along with all the rest of your genetic material. Now it exists, in duplicate form, all over your body. But when you in your turn come to have children, this chromosome will be destroyed when you manufacture eggs (or sperms). Bits of it will be interchanged with bits of your maternal chromosome number 8b. In any one sex cell, a new chromosome number 8 will be created, perhaps 'better' than the old one, perhaps 'worse', but, barring a rather improbable coincidence, definitely different, definitely unique. The life-span of a chromosome is one generation.

What about the life-span of a smaller genetic unit, say 1/100 of the length of your chromosome 8a? This unit too came from your father, but it very probably was not originally assembled in him. Following the earlier reasoning, there is a 99 per cent chance that he received it intact from one of his two parents. Suppose it was from his mother, your paternal grandmother. Again, there is a 99 per cent chance that she inherited it intact from one of her parents. Eventually, if we trace the ancestry of a small genetic unit back far enough, we will come to its original creator. At some stage it must have been created for the first time inside a testicle or an ovary of one of your ancestors.

Let me repeat the rather special sense in which I am using the word 'create'. The smaller sub-units which make up the genetic unit we are considering may well have existed long before. Our genetic unit was created at a particular moment only in the sense that the particular *arrangement* of sub-units by which it is defined did not exist before that moment. The moment of creation may have occurred quite recently, say in one of your grandparents. But if we consider a very small genetic unit, it may have been first assembled in a much more distant ancestor, perhaps an ape-like pre-human ancestor. Moreover, a small genetic unit inside you may go on just as far into the future, passing intact through a long line of your descendants.

Remember too that an individual's descendants constitute not a single line but a branching line. Whichever of your ancestors it was who 'created' a particular short length of your chromosome 8a, he or she very likely has many other descendants besides you. One of your genetic units may also be present in your second cousin. It may be present in me, and in the Prime Minister, and in your dog, for we all share ancestors if we go back far enough. Also the same small unit might be assembled several times independently by chance: if the unit is small, the coincidence is not too improbable. But even a close relative is unlikely to share a whole chromosome with you. The smaller a genetic unit is, the more likely it is that another individual shares it—the more likely it is to be represented many times over in the world, in the form of copies.

The chance coming together, through crossing-over, of previously existing sub-units is the usual way for a new genetic unit to be formed. Another way—of great evolutionary importance even though it is rare—is called *point mutation*. A point mutation is an error corresponding to a single misprinted letter in a book. It is rare, but clearly the longer a genetic unit is, the more likely it is to be altered by a mutation somewhere along its length.

Another rare kind of mistake or mutation which has important long-term consequences is called *inversion*. A piece of chromosome detaches itself at both ends, turns head over heels, and reattaches itself in the inverted position. In terms of the earlier analogy, this would necessitate some renumbering of pages. Sometimes portions of chromosomes do not simply invert, but become reattached in a completely different part of the chromosome, or even join up with a different chromosome altogether. This corresponds to the transfer of a wad of pages from one volume to another. The importance of this kind of mistake is that, though usually disastrous, it can occasionally lead to the close *linkage* of pieces of genetic material which happen to work well together. Perhaps two cistrons which have a beneficial effect only when they are both present—they complement or reinforce each other in some way—will be brought close to each other by means of inversion. Then natural selection may tend to favour the new 'genetic unit' so formed, and it will spread through the future population. It is possible that gene complexes have, over the years, been extensively rearranged or 'edited' in this kind of way.

One of the neatest examples of this concerns the phenomenon known as *mimicry*. Some butterflies taste nasty. They are usually brightly and distinctively coloured, and birds learn to avoid them by their 'warning' marks. Now other species of butterfly which do not taste nasty cash in. They *mimic* the nasty ones. They are born looking like them in colour and shape (but not taste). They frequently fool human naturalists, and they also fool birds. A bird who has once tasted a genuinely nasty butterfly tends to avoid all butterflies who look the same. This includes the mimics, and so genes for mimicry are favoured by natural selection. That is how mimicry evolves.

There are many different species of 'nasty' butterfly and they do not all look alike. A mimic cannot resemble all of them: it has to commit itself to one particular nasty species. In general, any particular species of mimic is a specialist at mimicking one particular nasty species. But there are species of mimic which do something very strange. Some individuals of the species mimic one nasty species; other individuals mimic another. Any individual who was intermediate or who tried to mimic both would soon be eaten; but such intermediates are not born. Just as an individual is either definitely male or definitely female, so an individual butterfly mimics either one nasty species or the other. One butterfly may mimic species *A* while his brother mimics species *B*.

It looks as though a single gene determines whether an individual will mimic species *A* or species *B*. But how can a single gene determine all the multifarious aspects of mimicry—colour, shape, spot pattern, rhythm of flight? The answer is that one gene in the sense of a *cistron* probably cannot. But by the unconscious and automatic 'editing' achieved by inversions and other accidental rearrangements of genetic material, a large cluster of formerly separate genes has come together in a tight linkage group on a chromosome. The whole cluster behaves like a single gene—indeed, by our definition it now *is* a single gene—and it has an 'allele' which is really another cluster. One cluster contains the cistrons concerned with mimicking species *A*; the other those concerned with mimicking species *B*. Each cluster is so rarely split up by crossing-over that an intermediate butterfly is never seen in nature, but they do very occasionally turn up if large numbers of butterflies are bred in the laboratory.

I am using the word gene to mean a genetic unit which is small enough to last for a large number of generations and to be distributed around in the form of many copies. This is not a rigid all-or-nothing definition, but a kind of fading-out definition, like the definition of 'big' or 'old'. The more likely a length of chromosome is to be split by crossing-over, or altered by mutations of various kinds, the less it qualifies to be called a gene in the sense in which I am using the term. A cistron presumably qualifies, but so also do larger units. A dozen cistrons may be so close to each other on a chromosome that for our purposes they constitute a single long-lived genetic unit. The butterfly mimicry cluster is a good example. As the cistrons leave one body and enter the next, as they board sperm or egg for the journey into the next generation, they are likely to find the little vessel contains their close neighbours of the previous voyage, old shipmates with whom they sailed on the long odyssey from the bodies of distant ancestors. Neighbouring cistrons on the same chromosome form a tightly-knit troupe of travelling companions who seldom fail to get on board the same vessel when meiosis time comes around.

To be strict, this book should be called not *The Selfish Cistron* nor *The Selfish Chromosome*, but *The slightly selfish big bit of chromosome and the even more selfish little bit of chromosome*. To say the least this is not a catchy title so, defining a gene as a little bit of chromosome which potentially lasts for many generations, I call the book *The Selfish Gene*.

We have now arrived back at the point we left at the end of Chapter 1. There we saw that selfishness is to be expected in any entity which deserves the title of a basic unit of natural selection. We saw that some people regard the species as the unit of natural selection, others the population or group within the species, and yet others the individual. I said that I preferred to think of the gene as the fundamental unit of natural selection, and therefore the fundamental unit of self-interest. What I have now done is to *define* the gene in such a way that I cannot really help being right!

Natural selection in its most general form means the differential survival of entities. Some entities live and others die but, in order for this selective death to have any impact on the world, an additional condition must be met. Each entity must exist in the form of lots of copies, and at least some of the entities must be *potentially* capable of surviving—in the form of copies—for a significant period of evolutionary time. Small genetic units have these properties; individuals, groups, and species do not. It was the great achievement of Gregor Mendel to show that hereditary units can be treated in practice as indivisible and independent particles. Nowadays we know that this is a little too simple. Even a cistron is occasionally divisible and any two genes on the same chromosome are not wholly independent. What I have done is to define a gene as a unit which, to a high degree, *approaches* the ideal of indivisible particulateness. A gene is not indivisible, but it is seldom divided. It is either definitely present or definitely absent in the body of any given individual. A gene travels intact from grandparent to grandchild, passing straight through the intermediate generation without being merged with other genes. If genes continually blended with each other, natural selection as we now understand it would be impossible. Incidentally, this was proved in Darwin's lifetime, and it caused Darwin great worry since in those days it was assumed that heredity was a blending process. Mendel's discovery had already been published, and it could have rescued Darwin, but alas he never knew about it: nobody seems to have read it until years after Darwin and Mendel had both died. Mendel perhaps did not realize the significance of his findings, otherwise he might have written to Darwin.

Another aspect of the particulateness of the gene is that it does not grow senile; it is no more likely to die when it is a million years old than when it is only a hundred. It leaps from body to body down the generations, manipulating body after body in its own way and for its own ends, abandoning a succession of mortal bodies before they sink in senility and death.

The genes are the immortals, or rather, they are defined as genetic entities which come close to deserving the title. We, the individual survival machines in the world, can expect to live a few more decades. But the genes in the world have an expectation of life which must be measured not in decades but in thousands and millions of years.

In sexually reproducing species, the individual is too large and too temporary a genetic unit to qualify as a significant unit of natural selection. The group of individuals is an even larger unit. Genetically speaking, individuals and groups are like clouds in the sky or dust-storms in the desert. They are temporary aggregations or federations. They are not stable through evolutionary time. Populations may last a long while, but they are constantly blending with other populations and so losing their identity. They are also subject to evolutionary change from within. A population is not a discrete enough entity to be a unit of natural selection, not stable and unitary enough to be 'selected' in preference to another population.

An individual body seems discrete enough while it lasts, but alas, how long is that? Each individual is unique. You cannot get evolution by selecting between entities when there is only one copy of each entity! Sexual reproduction is not replication. Just as a population is contaminated by other populations, so an individual's posterity is contaminated by that of his sexual partner. Your children are only half you, your grandchildren only a quarter you. In a few generations the most you can hope for is a large number of descendants, each of whom bears only a tiny portion of you—a few genes—even if a few do bear your surname as well.

Individuals are not stable things, they are fleeting. Chromosomes too are shuffled into oblivion, like hands of cards soon after they are dealt. But the cards themselves survive the shuffling. The cards are the genes. The genes are not destroyed by crossing-over, they merely change partners and march on. Of course they march on. That is their business. They are the replicators and we are their survival machines. When we have served our purpose we are cast aside. But genes are denizens of geological time: genes are forever.

Genes, like diamonds, are forever, but not quite in the same way as diamonds. It is an individual diamond crystal which lasts, as an unaltered pattern of atoms. DNA molecules don't have that kind of permanence. The life of any one physical DNA molecule is quite short—perhaps a matter of months, certainly not more than one lifetime. But a DNA molecule could theoretically live on in the form of *copies* of itself for a hundred million years. Moreover, just like the ancient replicators in the primeval soup, copies of a particular gene may be distributed all over the world. The difference is that the modern versions are all neatly packaged inside the bodies of survival machines.

What I am doing is emphasizing the potential near-immortality of a gene, in the form of copies, as its defining property. To define a gene as a single cistron is good for some purposes, but for the purposes of evolutionary theory it needs to be enlarged. The extent of the enlargement is determined by the purpose of the definition. We want to find the practical unit of natural selection. To do this we begin by identifying the properties which a successful unit of natural selection must have. In the terms of the last chapter, these are longevity, fecundity, and copying-fidelity. We then simply define a 'gene' as the largest entity which, at least potentially, has these properties. The gene is a long-lived replicator, existing in the form of many duplicate copies. It is not infinitely long-lived. Even a diamond is not literally everlasting, and even a cistron can be cut in two by crossing-over. The gene is defined as a piece of chromosome which is sufficiently short for it to last, potentially, for *long enough* for it to function as a significant unit of natural selection.

Exactly how long is 'long enough'? There is no hard and fast answer. It will depend on how severe the natural selection 'pressure' is. That is, on how much more likely a 'bad' genetic unit is to die than its 'good' allele. This is a matter of quantitative detail which will vary from example to example. The largest practical unit of natural selection—the gene—will usually be found to lie somewhere on the scale between cistron and chromosome.

It is its potential immortality that makes a gene a good candidate as the basic unit of natural selection. But now the time has come to stress the word 'potential'. A gene *can* live for a million years, but many new genes do not even make it past their first generation. The few new ones who succeed do so partly because they are lucky, but mainly because they have what it takes, and that means they are good at making survival machines. They have an effect on the embryonic development of each successive body in which they find themselves, such that that body is a little bit more likely to live and reproduce than it would have been under the influence of the rival gene or allele. For example, a 'good' gene might ensure its survival by tending to endow the successive bodies in which it finds itself with long legs, which help those bodies to escape from predators. This is a particular example, not a universal one. Long legs, after all, are not always an asset. To a mole they would be a handicap. Rather than bog ourselves down in details, can we think of any *universal* qualities which we would expect to find in all good (i.e. long-lived) genes? Conversely, what are the properties which instantly mark a gene out as a 'bad', short-lived one? There might be several such universal properties, but there is one which is particularly relevant to this book: at the gene level, altruism must be bad and selfishness good. This follows inexorably from our definitions of altruism and selfishness. Genes are competing directly with their alleles for survival, since their alleles in the gene pool are rivals for their slot on the chromosomes of future generations. Any gene which behaves in such a way as to increase its own survival chances in the gene pool at the expense of its alleles will, by definition, tautologously, tend to survive. The gene is the basic unit of selfishness.

The main message of this chapter has now been stated. But I have glossed over some complications and hidden assumptions. The first complication has already been briefly mentioned. However independent and free genes may be in their journey through the generations, they are very much *not* free and independent agents in their control of embryonic development. They collaborate and interact in inextricably complex ways, both with each other, and with their external environment. Expressions like 'gene for long legs' or 'gene for altruistic behaviour' are convenient figures of speech, but it is important to understand what they mean. There is no gene which single-handedly builds a leg, long or short. Building a leg is a multi-gene cooperative enterprise. Influences from the external environment too are indispensable: after all, legs are actually made of food! But there may well be a single gene which, *other things being equal*, tends to make legs longer than they would have been under the influence of the gene's allele.

As an analogy, think of the influence of a fertilizer, say nitrate, on the growth of wheat. Everybody knows that wheat plants grow bigger in the presence of nitrate than in its absence. But nobody would be so foolish as to claim that, on its own, nitrate can make a wheat plant. Seed, soil, sun, water, and various minerals are obviously all necessary as well. But if all these other factors are held constant, and even if they are allowed to vary within limits, addition of nitrate will make the wheat plants grow bigger. So it is with single genes in the development of an embryo. Embryonic development is controlled by an interlocking web of relationships so complex that we had best not contemplate it. No one factor, genetic or environmental, can be considered as the single 'cause' of any part of a baby. All parts of a baby have a near infinite number of antecedent causes. But a *difference* between one baby and another, for example a difference in length of leg, might easily be traced to one or a few simple antecedent differences, either in environment or in genes. It is *differences* which matter in the competitive struggle to survive; and it is genetically-controlled differences which matter in evolution.

As far as a gene is concerned, its alleles are its deadly rivals, but other genes are just a part of its environment, comparable to temperature, food, predators, or companions. The effect of the gene depends on its environment, and this includes other genes. Sometimes a gene has one effect in the presence of a particular other gene, and a completely different effect in the presence of another set of companion genes. The whole set of genes in a body constitutes a kind of genetic climate or background, modifying and influencing the effects of any particular gene.

But now we seem to have a paradox. If building a baby is such an intricate cooperative venture, and if every gene needs several thousands of fellow genes to complete its task, how can we reconcile this with my picture of indivisible genes, springing like immortal chamois from body to body down the ages: the free, untrammelled, and self-seeking agents of life? Was that all nonsense? Not at all. I may have got a bit carried away with the purple passages, but I was not talking nonsense, and there is no real paradox. We can explain this by means of another analogy.

One oarsman on his own cannot win the Oxford and Cambridge boat race. He needs eight colleagues. Each one is a specialist who always sits in a particular part of the boat—bow or stroke or cox etc. Rowing the boat is a cooperative venture, but some men are nevertheless better at it than others. Suppose a coach has to choose his ideal crew from a pool of candidates, some specializing in the bow position, others specializing as cox, and so on. Suppose that he makes his selection as follows. Every day he puts together three new trial crews, by random shuffling of the candidates for each position, and he makes the three crews race against each other. After some weeks of this it will start to emerge that the winning boat often tends to contain the same individual men. These are marked up as good oarsmen. Other individuals seem consistently to be found in slower crews, and these are eventually rejected. But even an outstandingly good oarsman might sometimes be a member of a slow crew, either because of the inferiority of the other members, or because of bad luck—say a strong adverse wind. It is only *on average* that the best men tend to be in the winning boat.

The oarsmen are genes. The rivals for each seat in the boat are alleles potentially capable of occupying the same slot along the length of a chromosome. Rowing fast corresponds to building a body which is successful at surviving. The wind is the external environment. The pool of alternative candidates is the gene pool. As far as the survival of any one body is concerned, all its genes are in the same boat. Many a good gene gets into bad company, and finds itself sharing a body with a lethal gene, which kills the body off in childhood. Then the good gene is destroyed along with the rest. But this is only one body, and replicas of the same good gene live on in other bodies which lack the lethal gene. Many copies of good genes are dragged under because they happen to share a body with bad genes, and many perish through other forms of ill luck, say when their body is struck by lightning. But by definition luck, good and bad, strikes at random, and a gene which is *consistently* on the losing side is not unlucky; it is a bad gene.

One of the qualities of a good oarsman is teamwork, the ability to fit in and cooperate with the rest of a crew. This may be just as important as strong muscles. As we saw in the case of the butter-flies, natural selection may unconsciously 'edit' a gene complex by means of inversions and other gross movements of bits of chromosome, thereby bringing genes which cooperate well together into closely linked groups. But there is also a sense in which genes which are in no way linked to each other physically can be selected for their mutual compatibility. A gene which cooperates well with most of the other genes which it is likely to meet in successive bodies, i.e. the genes in the whole of the rest of the gene pool, will tend to have an advantage.

For example, a number of attributes are desirable in an efficient carnivore's body, among them sharp cutting teeth, the right kind of intestine for digesting meat, and many other things. An efficient herbivore, on the other hand, needs flat grinding teeth, and a much longer intestine with a different kind of digestive chemistry. In a herbivore gene pool, any new gene which conferred on its possessors sharp meat-eating teeth would not be very successful. This is not because meat-eating is universally a bad idea, but because you cannot efficiently eat meat unless you also have the right sort of intestine, and all the other attributes of a meat-eating way of life. Genes for sharp, meat-eating teeth are not inherently bad genes. They are only bad genes in a gene-pool which is dominated by genes for herbivorous qualities.

This is a subtle, complicated idea. It is complicated because the 'environment' of a gene consists largely of other genes, each of which is itself being selected for its ability to cooperate with *its* environment of other genes. An analogy adequate to cope with this subtle point does exist, but it is not from everyday experience. It is the analogy with human 'game theory', which will be introduced in Chapter 5 in connection with aggressive contests between individual animals. I therefore postpone further discussion of this point until the end of that chapter, and return to the central message of this one. This is that the basic unit of natural selection is best regarded not as the species, nor as the population, nor even as the individual, but as some small unit of genetic material which it is convenient to label the gene. The cornerstone of the argument, as given earlier, was the assumption that genes are potentially immortal, while bodies and all other higher units are temporary. This assumption rests upon two facts: the fact of sexual reproduction and crossing-over, and the fact of individual mortality. These facts are undeniably true. But this does not stop us asking why they are true. Why do we and most other survival machines practise sexual reproduction? Why do our chromosomes cross over? Any why do we not live for ever?

The question of why we die of old age is a complex one, and the details are beyond the scope of this book. In addition to particular reasons, some more general ones have been proposed. For example, one theory is that senility represents an accumulation of deleterious copying errors and other kinds of gene damage which occur during the individual's lifetime. Another theory, due to Sir Peter Medawar, is a good example of evolutionary thinking in terms of gene selection. Medawar first dismisses traditional arguments such as: 'Old individuals die as an act of altruism to the rest of the species, because if they stayed around when they were too decrepit to reproduce, they would clutter up the world to no good purpose.' As Medawar points out, this is a circular argument, assuming what it sets out to prove, namely that old animals are too decrepit to reproduce. It is also a naïve group-selection or species-selection kind of explanation, although that part of it could be rephrased more respectably. Medawar's own theory has a beautiful logic. We can build up to it as follows.

We have already asked what are the most general attributes of a 'good' gene, and we decided that 'selfishness' was one of them. But another general quality that successful genes will have is a tendency to postpone the death of their survival machines at least until after reproduction. No doubt some of your cousins and great-uncles died in childhood, but not a single one of your ancestors did. Ancestors just don't die young!

A gene which makes its possessors die is called a lethal gene. A semi-lethal gene has some debilitating effect, such that it makes death from other causes more probable. Any gene exerts its maximum effect on bodies at some particular stage of life, and lethals and semi-lethals are not exceptions. Most genes exert their influence during foetal life, others during childhood, others during young adulthood, others in middle age, and yet others in old age. (Reflect that a caterpillar and the butterfly it turns into have exactly the same set of genes.) Obviously lethal genes will tend to be removed from the gene pool. But equally obviously a late-acting lethal will be more stable in the gene pool than an early-acting lethal. A gene which is lethal in an older body may still be successful in the gene pool, provided its lethal effect does not show itself until after the body has had time to do at least some reproducing. For instance, a gene which made old bodies develop cancer could be passed on to numerous offspring because the individuals would reproduce before they got cancer. On the other hand, a gene which made young adult bodies develop cancer would not be passed on to very many offspring, and a gene which made young children develop fatal cancer would not be passed on to any offspring at all. According to this theory then, senile decay is simply a by-product of the accumulation in the gene pool of late-acting lethal and semi-lethal genes, which have been allowed to slip through the net of natural selection simply because they are late-acting.

The aspect that Medawar himself emphasizes is that selection will favour genes which have the effect of postponing the operation of other, lethal genes, and it will also favour genes which have the effect of hastening the effect of good genes. It may be that a great deal of evolution consists of genetically-controlled changes in the time of onset of gene activity.

It is important to notice that this theory does not need to make any prior assumptions about reproduction occurring only at certain ages. Taking as a starting assumption that all individuals were equally likely to have a child at any age, the Medawar theory would quickly predict the accumulation in the gene pool of late-acting deleterious genes, and the tendency to reproduce less in old age would follow as a secondary consequence.

As an aside, one of the good features of this theory is that it leads us to some rather interesting speculations. For instance it follows from it that if we wanted to increase the human life span, there are two general ways in which we could do it. Firstly, we could ban reproduction before a certain age, say forty. After some centuries of this the minimum age limit would be raised to fifty, and so on. It is conceivable that human longevity could be pushed up to several centuries by this means. I cannot imagine that anyone would seriously want to institute such a policy.

Secondly we could try to 'fool' genes into thinking that the body they are sitting in is younger than it really is. In practice this would mean identifying changes in the internal chemical environment of a body which take place during ageing. Any of these could be the 'cues' which 'turn on' late-acting lethal genes. By simulating the superficial chemical properties of a young body it might be possible to prevent the turning on of late-acting deleterious genes. The interesting point is that chemical signals of old age need not in any normal sense be deleterious in themselves. For instance, suppose that it incidentally happens to be a fact that a substance S is more concentrated in the bodies of old individuals than of young individuals. S in itself might be quite harmless, perhaps some substance in the food which accumulates in the body over time. But automatically, any gene which just happened to exert a deleterious effect in the presence of S, but which otherwise had a good effect, would be positively selected in the gene pool, and would in effect *be* a gene 'for' dying of old age. The cure would simply be to remove S from the body.

What is revolutionary about this idea is that S itself is only a 'label' for old age. Any doctor who noticed that high concentrations of S tended to lead to death, would probably think of S as a kind of poison, and would rack his brains to find a direct causal link between *S* and bodily malfunctioning. But in the case of our hypothetical example, he might be wasting his time!

There might also be a substance *Y*, a 'label' for youth in the sense that it was more concentrated in young bodies than in old ones. Once again, genes might be selected which would have good effects in the presence of *Y*, but which would be deleterious in its absence. Without having any way of knowing what *S* or *Y* are—there could be many such substances—we can simply make the general prediction that the more you can simulate or mimic the properties of a young body in an old one, however superficial these properties may seem, the longer should that old body live.

I must emphasize that these are just speculations based on the Medawar theory. Although there is a sense in which the Medawar theory logically must have some truth in it, this does not mean necessarily that it is the right explanation for any given practical example of senile decay. What matters for present purposes is that the gene-selection view of evolution has no difficulty in accounting for the tendency of individuals to die when they get old. The assumption of individual mortality, which lay at the heart of our argument in this chapter, is justifiable within the framework of the theory.

The other assumption I have glossed over, that of the existence of sexual reproduction and crossing-over, is more difficult to justify. Crossing-over does not always have to happen. Male fruit-flies do not do it. There is a gene which has the effect of suppressing crossing-over in females as well. If we were to breed a population of flies in which this gene was universal, the *chromosome* in a 'chromosome pool' would become the basic indivisible unit of natural selection. In fact, if we followed our definition to its logical conclusion, a whole chromosome would have to be regarded as one 'gene'.

Then again, alternatives to sex do exist. Female greenflies can bear live, fatherless, female offspring, each one containing all the genes of its mother. (Incidentally, an embryo in her mother's 'womb' may have an even smaller embryo inside her own womb. So a greenfly female may give birth to a daughter and a granddaughter simultaneously, both of them being equivalent to her own identical twins). Many plants propagate vegetatively by sending out suckers. In this case we might prefer to speak of *growth* rather than of reproduction; but then, if you think about it, there is rather little distinction between growth and non-sexual reproduction anyway, since both occur by simple mitotic cell division. Sometimes the plants produced by vegetative reproduction become detached from the 'parent'. In other cases, for instance elm trees, the connecting suckers remain intact. In fact an entire elm wood might be regarded as a single individual.

So, the question is: if greenflies and elm trees don't do it, why do the rest of us go to such lengths to mix our genes up with somebody else's before we make a baby? It does seem an odd way to proceed. Why did sex, that bizarre perversion of straightforward replication, ever arise in the first place? What is the good of sex?

This is an extremely difficult question for the evolutionist to answer. Most serious attempts to answer it involve sophisticated mathematical reasoning. I am frankly going to evade it except to say one thing. This is that at least some of the difficulty which theorists have with explaining the evolution of sex results from the fact that they habitually think of the individual as trying to maximize the number of his genes which survive. In these terms, sex appears paradoxical because it is an 'inefficient' way for an individual to propagate his genes: each child has only 50 per cent of the individual's genes, the other 50 per cent being provided by the sexual partner. If only, like a greenfly, he would bud-off children who were exact replicas of himself, he would pass 100 per cent of his genes on to the next generation in the body of every child. This apparent paradox has driven some theorists to embrace group-selectionism, since it is relatively easy to think of group-level advantages for sex. As W. F. Bodmer has succinctly put it, sex 'facilitates the accumulation in a single individual of advantageous mutations which arose separately in different individuals.'

But the paradox seems less paradoxical if we follow the argument of this book, and treat the individual as a survival machine built by a short-lived confederation of long-lived genes. 'Efficiency' from the whole individual's point of view is then seen to be irrelevant. Sexuality versus non-sexuality will be regarded as an attribute under single-gene control, just like blue eyes versus brown eyes. A gene 'for' sexuality manipulates all the other genes for its own selfish ends. So does a gene for crossing-over. There are even genes—called mutators—which manipulate the rates of copying-errors in other genes. By definition, a copying error is to the disadvantage of the gene which is miscopied. But if it is to the advantage of the selfish mutator gene which induces it, the mutator can spread through the gene pool. Similarly, if crossing-over benefits a gene for crossing-over, that is a sufficient explanation for the existence of crossing-over. And if sexual, as opposed to non-sexual, reproduction benefits a gene for sexual reproduction, that is a sufficient explanation for the existence of sexual reproduction. Whether or not it benefits all the rest of an individual's genes is comparatively irrelevant. Seen from the selfish gene's point of view, sex is not so bizarre after all.

This comes perilously close to being a circular argument, since the existence of sexuality is a precondition for the whole chain of reasoning which leads to the gene being regarded as the unit of selection. I believe there are ways of escaping from the circularity, but this book is not the place to pursue the question. Sex exists. That much is true. It is a consequence of sex and crossing-over that the small genetic unit or gene can be regarded as the nearest thing we have to a fundamental, independent agent of evolution.

Sex is not the only apparent paradox which becomes less puzzling the moment we learn to think in selfish gene terms. For instance, it appears that the amount of DNA in organisms is more than is strictly necessary for building them: a large fraction of the DNA is never translated into protein. From the point of view of the individual organism this seems paradoxical. If the 'purpose' of DNA is to supervise the building of bodies, it is surprising to find a large quantity of DNA which does no such thing. Biologists are racking their brains trying to think what useful task this apparently surplus DNA is doing. But from the point of view of the selfish genes themselves, there is no paradox. The true 'purpose' of DNA is to survive, no more and no less. The simplest way to explain the surplus DNA is to suppose that it is a parasite, or at best a harmless but useless passenger, hitching a ride in the survival machines created by the other DNA.

Some people object to what they see as an excessively gene-centred view of evolution. After all, they argue, it is whole individuals with all their genes who actually live or die. I hope I have said enough in this chapter to show that there is really no disagreement here. Just as whole boats win or lose races, it is indeed individuals who live or die, and the *immediate* manifestation of natural selection is nearly always at the individual level. But the long-term consequences of non-random individual death and reproductive success are manifested in the form of changing gene frequencies in the gene pool. With reservations, the gene pool plays the same role for the modern replicators as the primeval soup did for the original ones. Sex and chromosomal crossing-over have the effect of preserving the liquidity of the modern equivalent of the soup. Because of sex and crossing-over the gene pool is kept well stirred, and the genes partially shuffled. Evolution is the process by which some genes become more numerous and others less numerous in the gene pool. It is good to get into the habit, whenever we are trying to explain the evolution of some characteristic, such as altruistic behaviour, of asking ourselves simply: 'what effect will this characteristic have on frequencies of genes in the gene pool?' At times, gene language gets a bit tedious, and for brevity and vividness we shall lapse into metaphor. But we shall always keep a sceptical eye on our metaphors, to make sure they can be translated back into gene language if necessary.

As far as the gene is concerned, the gene pool is just the new sort of soup where it makes its living. All that has changed is that nowadays it makes its living by cooperating with successive groups of companions drawn from the gene pool in building one mortal survival machine after another. It is to survival machines themselves, and the sense in which genes may be said to control their behaviour, that we turn in the next chapter.

4. The gene machine

SURVIVAL machines began as passive receptacles for the genes, providing little more than walls to protect them from the chemical warfare of their rivals and the ravages of accidental molecular bombardment. In the early days they 'fed' on organic molecules freely available in the soup. This easy life came to an end when the organic food in the soup, which had been slowly built up under the energetic influence of centuries of sunlight, was all used up. A major branch of survival machines, now called plants, started to use sunlight directly themselves to build up complex molecules from simple ones, re-enacting at much higher speed the synthetic processes of the original soup. Another branch, now known as animals, 'discovered' how to exploit the chemical labours of the plants, either by eating them, or by eating other animals. Both main branches of survival machines evolved more and more ingenious tricks to increase their efficiency in their various ways of life, and new ways of life were continually being opened up. Sub-branches and sub-sub-branches evolved, each one excelling in a particular specialized way of making a living: in the sea, on the ground, in the air, underground, up trees, inside other living bodies. This sub-branching has given rise to the immense diversity of animals and plants which so impresses us today.

Both animals and plants evolved into many-celled bodies, complete copies of all the genes being distributed to every cell. We do not know when, why, or how many times independently, this happened. Some people use the metaphor of a colony, describing a body as a colony of cells. I prefer to think of the body as a colony of *genes*, and of the cell as a convenient working unit for the chemical industries of the genes.

Colonies of genes they may be but, in their behaviour, bodies have undeniably acquired an individuality of their own. An animal moves as a coordinated whole, as a unit. Subjectively I feel like a unit, not a colony. This is to be expected. Selection has favoured genes which cooperate with others. In the fierce competition for scarce resources, in the relentless struggle to eat other survival machines, and to avoid being eaten, there must have been a premium on central coordination rather than anarchy within the communal body. Nowadays the intricate mutual co-evolution of genes has proceeded to such an extent that the communal nature of an individual survival machine is virtually unrecognizable. Indeed many biologists do not recognize it, and will disagree with me.

Fortunately for what journalists would call the 'credibility' of the rest of this book, the disagreement is largely academic. Just as it is not convenient to talk about quanta and fundamental particles when we discuss the workings of a car, so it is often tedious and unnecessary to keep dragging genes in when we discuss the behaviour of survival machines. In practice it is usually convenient, as an approximation, to regard the individual body as an agent 'trying' to increase the numbers of all its genes in future generations. I shall use the language of convenience. Unless otherwise stated, 'altruistic behaviour' and 'selfish behaviour' will mean behaviour directed by one animal body toward another.

This chapter is about *behaviour*—the trick of rapid movement which has been largely exploited by the animal branch of survival machines. Animals became active go-getting gene vehicles: gene machines. The characteristic of behaviour, as biologists use the term, is that it is fast. Plants move, but very slowly. When seen in highly speeded-up film, climbing plants look like active animals. But most plant movement is really irreversible growth. Animals, on the other hand, have evolved ways of moving hundreds of thousands of times faster. Moreover, the movements they make are reversible, and repeatable an indefinite number of times.

The gadget which animals evolved to achieve rapid movement was the muscle. Muscles are engines which, like the steam engine and the internal combustion engine, use energy stored in chemical fuel to generate mechanical movement. The difference is that the immediate mechanical force of a muscle is generated in the form of tension, rather than gas pressure as in the case of the steam and internal combustion engines. But muscles are like engines in that they often exert their force on cords, and levers with hinges. In us the levers are known as bones, the cords as tendons, and the hinges as joints. Quite a lot is known about the exact molecular ways in which muscles work, but I find more interesting the question of how muscle contractions are *timed*.

Have you ever watched an artificial machine of some complexity, a knitting or sewing machine, a loom, an automatic bottling factory, or a hay baler? Motive power comes from somewhere, an electric motor say, or a tractor. But much more baffling is the intricate timing of the operations. Valves open and shut in the right order, steel fingers deftly tie a knot round a hay bale, and then at just the right moment a knife shoots out and cuts the string. In many artificial machines timing is achieved by that brilliant invention the cam. This translates simple rotary motion into a complex rhythmic pattern of operations by means of an eccentric or specially shaped wheel. The principle of the musical box is similar. Other machines such as the steam organ and the pianola use paper rolls or cards with holes punched in a pattern. Recently there has been a trend towards replacing such simple mechanical timers with electronic ones. Digital computers are examples of large and versatile electronic devices which can be used for generating complex timed patterns of movements. The basic component of a modern electronic machine like a computer is the semiconductor, of which a familiar form is the transistor.

Survival machines seem to have bypassed the cam and the punched card altogether. The apparatus they use for timing their movements has more in common with an electronic computer, although it is strictly different in fundamental operation. The basic unit of biological computers, the nerve cell or neurone, is really nothing like a transistor in its internal workings. Certainly the code in which neurones communicate with each other seems to be a little bit like the pulse codes of digital computers, but the individual neurone is a much more sophisticated data-processing unit than the transistor. Instead of just three connections with other components, a single neurone may have tens of thousands. The neurone is slower than the transistor, but it has gone much further in the direction of miniaturization, a trend which has dominated the electronics industry over the past two decades. This is brought home by the fact that there are some ten thousand million neurones in the human brain: you could pack only a few hundred transistors into a skull.

Plants have no need of the neurone, because they get their living without moving around, but it is found in the great majority of animal groups. It may have been 'discovered' early in animal evolution, and inherited by all groups, or it may have been rediscovered several times independently.

Neurones are basically just cells, with a nucleus and chromosomes like other cells. But their cell walls are drawn out in long, thin, wire-like projections. Often a neurone has one particularly long 'wire' called the axon. Although the width of an axon is microscopic, its length may be many feet: there are single axons which run the whole length of a giraffe's neck. The axons are usually bundled together in thick multi-stranded cables called nerves. These lead from one part of the body to another carrying messages, rather like trunk telephone cables. Other neurones have short axons, and are confined to dense concentrations of nervous tissue called ganglia, or, when they are very large, brains. Brains may be regarded as analogous in function to computers. They are analogous in that both types of machine generate complex patterns of output, after analysis of complex patterns of input, and after reference to stored information.

The main way in which brains actually contribute to the success of survival machines is by controlling and coordinating the contractions of muscles. To do this they need cables leading to the muscles, and these are called motor nerves. But this leads to efficient preservation of genes only if the timing of muscle contractions bears some relation to the timing of events in the outside world. It is important to contract the jaw muscles only when the jaws contain something worth biting, and to contract the leg muscles in running patterns only when there is something worth running towards or away from. For this reason, natural selection favoured animals which became equipped with sense organs, devices which translate patterns of physical events in the outside world into the pulse code of the neurones. The brain is connected to the sense organs—eyes, ears, taste-buds etc.—by means of cables called sensory nerves. The workings of the sensory systems are particularly baffling, because they can achieve far more sophisticated feats of pattern-recognition than the best and most expensive man-made machines; if this were not so, all typists would be redundant, superseded by speech-recognizing machines, or machines for reading handwriting. Human typists will be needed for many decades yet.

There may have been a time when sense organs communicated more or less directly with muscles; indeed, sea anemones are not far from this state today, since for their way of life it is efficient. But to achieve more complex and indirect relationships between the timing of events in the outside world and the timing of muscular contractions, some kind of brain was needed as an intermediary. A notable advance was the evolutionary 'invention' of memory. By this device, the timing of muscle contractions could be influenced not only by events in the immediate past, but by events in the distant past as well. The memory, or store, is an essential part of a digital computer too. Computer memories are more reliable than human ones, but they are less capacious, and enormously less sophisticated in their techniques of information-retrieval.

One of the most striking properties of survival-machine behaviour is its apparent purposiveness. By this I do not just mean that it seems to be well calculated to help the animal's genes to survive, although of course it is. I am talking about a closer analogy to human purposeful behaviour. When we watch an animal 'searching' for food, or for a mate, or for a lost child, we can hardly help imputing to it some of the subjective feelings we ourselves experience when we search. These may include 'desire' for some object, a 'mental picture' of the desired object, an 'aim' or 'end in view'. Each one of us knows, from the evidence of his own introspection, that, at least in one modern survival machine, this purposiveness has evolved the property we call 'consciousness'. I am not philosopher enough to discuss what this means, but fortunately it does not matter for our present purposes because it is easy to talk about machines which behave *as if* motivated by a purpose, and to leave open the question whether they actually are conscious. These machines are basically very simple, and the principles of unconscious purposive behaviour are among the commonplaces of engineering science. The classic example is the Watt steam governor.

The fundamental principle involved is called negative feedback, of which there are various different forms. In general what happens is this. The 'purpose machine', the machine or thing that behaves as if it had a conscious purpose, is equipped with some kind of measuring device which measures the discrepancy between the current state of things, and the 'desired' state. It is built in such a way that the larger this discrepancy is, the harder the machine works. In this way the machine will automatically tend to reduce the discrepancy—this is why it is called *negative* feedback—and it may actually come to rest if the 'desired' state is reached. The Watt governor consists of a pair of balls which are whirled round by a steam engine. Each ball is on the end of a hinged arm. The faster the balls fly round, the more does centrifugal force push the arms towards a horizontal position, this tendency being resisted by gravity. The arms are connected to the steam valve feeding the engine, in such a way that the steam tends to be shut off when the arms approach the horizontal position. So, if the engine goes too fast, some of its steam will be shut off, and it will tend to slow down. If it slows down too much, more steam will automatically be fed to it by the valve, and it will speed up again. Such purpose machines often oscillate due to over-shooting and time-lags, and it is part of the engineer's art to build in supplementary devices to reduce the oscillations.

The 'desired' state of the Watt governor is a particular speed of rotation. Obviously it does not consciously desire it. The 'goal' of a machine is simply defined as that state to which it tends to return. Modern purpose machines use extensions of basic principles like negative feedback to achieve much more complex 'life-like' behaviour. Guided missiles, for example, appear to search actively for their target, and when they have it in range they seem to pursue it, taking account of its evasive twists and turns, and sometimes even 'predicting' or 'anticipating' them. The details of how this is done are not worth going into. They involve negative feedback of various kinds, 'feed-forward', and other principles well understood by engineers and now known to be extensively involved in the working of living bodies. Nothing remotely approaching consciousness needs to be postulated, even though a layman, watching its apparently deliberate and purposeful behaviour, finds it hard to believe that the missile is not under the direct control of a human pilot.

It is a common misconception that because a machine such as a guided missile was originally designed and built by conscious man, then it must be truly under the immediate control of conscious man. Another variant of this fallacy is 'computers do not really play chess, because they can only do what a human operator tells them'. It is important that we understand why this is fallacious, because it affects our understanding of the sense in which genes can be said to 'control' behaviour. Computer chess is quite a good example for making the point, so I will discuss it briefly.

Computers do not yet play chess as well as human grand masters, but they have reached the standard of a good amateur. More strictly, one should say *programs* have reached the standard of a good amateur, for a chess-playing program is not fussy which physical computer it uses to act out its skills. Now, what is the role of the human programmer? First, he is definitely not manipulating the computer from moment to moment, like a puppeteer pulling strings. That would be just cheating. He writes the program, puts it in the computer, and then the computer is on its own: there is no further human intervention, except for the opponent typing in his moves. Does the programmer perhaps anticipate all possible chess positions, and provide the computer with a long list of good moves, one for each possible contingency? Most certainly not, because the number of possible positions in chess is so great that the world would come to an end before the list had been completed. For the same reason, the computer cannot possibly be programmed to try out 'in its head' all possible moves, and all possible follow-ups, until it finds a winning strategy. There are more possible games of chess than there are atoms in the galaxy. So much for the trivial non-solutions to the problem of programming a computer to play chess. It is in fact an exceedingly difficult problem, and it is hardly surprising that the best programs have still not achieved grand master status.

The programmer's actual role is rather more like that of a father teaching his son to play chess. He tells the computer the basic moves of the game, not separately for every possible starting position, but in terms of more economically expressed rules. He does not literally say in plain English 'bishops move in a diagonal', but he does say something mathematically equivalent, such as, though more briefly: 'New coordinates of bishop are obtained from old coordinates, by adding the same constant, though not necessarily with the same sign, to both old x coordinate and old y coordinate.' Then he might program in some 'advice', written in the same sort of mathematical or logical language, but amounting in human terms to hints such as 'don't leave your king unguarded', or useful tricks such as 'forking' with the knight. The details are intriguing, but they would take us too far afield. The important point is this. When it is actually playing, the computer is on its own, and can expect no help from its master. All the programmer can do is to set the computer up *beforehand* in the best way possible, with a proper balance between lists of specific knowledge, and hints about strategies and techniques.

The genes too control the behaviour of their survival machines, not directly with their fingers on puppet strings, but indirectly like the computer programmer. All they can do is to set it up beforehand; then the survival machine is on its own, and the genes can only sit passively inside. Why are they so passive? Why don't they grab the reins and take charge from moment to moment? The answer is that they cannot because of time-lag problems. This is best shown by another analogy, taken from science fiction. *A for Andromeda* by Fred Hoyle and John Elliot is an exciting story, and, like all good science fiction, it has some interesting scientific points lying behind it. Strangely, the book seems to lack explicit mention of the most important of these underlying points. It is left to the reader's imagination. I hope the authors will not mind if I spell it out here.

There is a civilization 200 light years away, in the constellation of Andromeda. They want to spread their culture to distant worlds. How best to do it? Direct travel is out of the question. The speed of light imposes a theoretical upper limit to the rate at which you can get from one place to another in the universe, and mechanical considerations impose a much lower limit in practice. Besides, there may not be all that many worlds worth going to, and how do you know which direction to go in? Radio is a better way of communicating with the rest of the universe, since, if you have enough power to broadcast your signals in all directions rather than beam them in one direction, you can reach a very large number of worlds (the number increasing as the square of the distance the signal travels). Radio waves travel at the speed of light, which means the signal takes 200 years to reach earth from Andromeda. The trouble with this sort of distance is that you can never hold a conversation. Even if you discount the fact that each successive message from earth would be transmitted by people separated from each other by twelve generations, it would be just plain wasteful to attempt to converse over such distances.

This problem will soon arise in earnest for us: it takes about four minutes for radio waves to travel between earth and Mars. There can be no doubt that spacemen will have to get out of the habit of conversing in short alternating sentences, and will have to use long soliloquies or monologues, more like letters than conversations. As another example, Roger Payne has pointed out that the acoustics of the sea have certain peculiar properties, which mean that the exceedingly loud 'song' of the humpback whale could theoretically be heard all the way round the world, provided the whales swim at a certain depth. It is not known whether they actually do communicate with each other over very great distances, but if they do they must be in much the same predicament as an astronaut on Mars. The speed of sound in water is such that it would take nearly two hours for the song to travel across the Atlantic Ocean and for a reply to return. I suggest this as an explanation for the fact that the whales deliver a continuous soliloquy, without repeating themselves, for a full eight minutes. They then go back to the beginning of the song and repeat it all over again, many times over, each complete cycle lasting about eight minutes.

The Andromedans of the story did the same thing. Since there was no point in waiting for a reply, they assembled everything they wanted to say into one huge unbroken message, and then they broadcast it out into space, over and over again, with a cycle time of several months. Their message was very different from that of the whales, however. It consisted of coded instructions for the building and programming of a giant computer. Of course the instructions were in no human language, but almost any code can be broken by a skilled cryptographer, especially if the designers of the code intended it to be easily broken. Picked up by the Jodrell Bank radio telescope, the message was eventually decoded, the computer built, and the program run. The results were nearly disastrous for mankind, for the intentions of the Andromedans were not universally altruistic, and the computer was well on the way to dictatorship over the world before the hero eventually finished it off with an axe.

From our point of view, the interesting question is in what sense the Andromedans could be said to be manipulating events on Earth. They had no direct control over what the computer did from moment to moment; indeed they had no possible way of even knowing the computer had been built, since the information would have taken 200 years to get back to them. The decisions and actions of the computer were entirely its own. It could not even refer back to its masters for general policy instructions. All its instructions had to be built-in in advance, because of the inviolable 200 year barrier. In principle, it must have been programmed very much like a chess-playing computer, but with greater flexibility and capacity for absorbing local information. This was because the program had to be designed to work not just on earth, but on any world possessing an advanced technology, any of a set of worlds whose detailed conditions the Andromedans had no way of knowing.

Just as the Andromedans had to have a computer on earth to take day-to-day decisions for them, our genes have to build a brain. But the genes are not only the Andromedans who sent the coded instructions; they are also the instructions themselves. The reason why they cannot manipulate our puppet strings directly is the same: time-lags. Genes work by controlling protein synthesis. This is a powerful way of manipulating the world, but it is slow. It takes months of patiently pulling protein strings to build an embryo. The whole point about behaviour, on the other hand, is that it is fast. It works on a time-scale not of months but of seconds and fractions of seconds. Something happens in the world, an owl flashes overhead, a rustle in the long grass betrays prey, and in milliseconds nervous systems crackle into action, muscles leap, and someone's life is saved—or lost. Genes don't have reaction-times like that. Like the Andromedans, the genes can only do their best *in advance* by building a fast executive computer for themselves, and programming it in advance with rules and 'advice' to cope with as many eventualities as they can 'anticipate'. But life, like the game of chess, offers too many different possible eventualities for all of them to be anticipated.

Like the chess programmer, the genes have to 'instruct' their survival machines not in specifics, but in the general strategies and tricks of the living trade.

As J. Z. Young has pointed out, the genes have to perform a task analogous to prediction. When an embryo survival machine is being built, the dangers and problems of its life lie in the future. Who can say what carnivores crouch waiting for it behind what bushes, or what fleet-footed prey will dart and zig-zag across its path? No human prophet, nor any gene. But some general predictions can be made. Polar bear genes can safely predict that the future of their unborn survival machine is going to be a cold one. They do not think of it as a prophecy, they do not think at all: they just build in a thick coat of hair, because that is what they have always done before in previous bodies, and that is why they still exist in the gene pool. They also predict that the ground is going to be snowy, and their prediction takes the form of making the coat of hair white and therefore camouflaged. If the climate of the Arctic changed so rapidly that the baby bear found itself born into a tropical desert, the predictions of the genes would be wrong, and they would pay the penalty. The young bear would die, and they inside it.

Prediction in a complex world is a chancy business. Every decision that a survival machine takes is a gamble, and it is the business of genes to program brains in advance so that on average they take decisions which pay off. The currency used in the casino of evolution is survival, strictly gene survival, but for many purposes individual survival is a reasonable approximation. If you go down to the water-hole to drink, you increase your risk of being eaten by predators who make their living lurking for prey by water-holes. If you do not go down to the water-hole you will eventually die of thirst. There are risks whichever way you turn, and you must take the decision which maximizes the long-term survival chances of your genes. Perhaps the best policy is to postpone drinking until you are very thirsty, then go and have one good long drink to last you a long time. That way you reduce the number of separate visits to the water-hole, but you have to spend a long time with your head down when you finally do drink. Alternatively the best gamble might be to drink little and often, snatching quick gulps of water while running past the water-hole. Which is the best gambling strategy depends on all sorts of complex things, not least the hunting habit of the predators, which itself is evolved to be maximally efficient from their point of view. Some form of weighing up of the odds has to be done. But of course we do not have to think of the animals as making the calculations consciously. All we have to believe is that those individuals whose genes build brains in such a way that they tend to gamble correctly are as a direct result more likely to survive, and therefore to propagate those same genes.

We can carry the metaphor of gambling a little further. A gambler must think of three main quantities, stake, odds, and prize. If the prize is very large, a gambler is prepared to risk a big stake. A gambler who risks his all on a single throw stands to gain a great deal. He also stands to lose a great deal, but on average high-stake gamblers are no better and no worse off than other players who play for low winnings with low stakes. An analogous comparison is that between speculative and safe investors on the stock market. In some ways the stock market is a better analogy than a casino, because casinos are deliberately rigged in the bank's favour (which means, strictly, that high-stake players will on average end up poorer than low-stake players; and low stake players poorer than those who do not gamble at all. But this is for a reason not germane to our discussion). Ignoring this, both high-stake play and low-stake play seem reasonable. Are there animal gamblers who play for high stakes, and others with a more conservative game? In Chapter 9 we shall see that it is often possible to picture males as high-stake high-risk gamblers, and females as safe investors, especially in polygamous species in which males compete for females. Naturalists who read this book may be able to think of species which can be described as high-stake high-risk players, and other species which play a more conservative game. I now return to the more general theme of how genes make 'predictions' about the future.

One way for genes to solve the problem of making predictions in rather unpredictable environments is to build in a capacity for learning. Here the program may take the form of the following instructions to the survival machine: 'Here is a list of things defined as rewarding: sweet taste in the mouth, orgasm, mild temperature, smiling child. And here is a list of nasty things:

various sorts of pain, nausea, empty stomach, screaming child. If you should happen to do something which is followed by one of the nasty things, don't do it again, but on the other hand repeat anything which is followed by one of the nice things.' The advantage of this sort of programming is that it greatly cuts down the number of detailed rules which have to be built into the original program; and it is also capable of coping with changes in the environment which could not have been predicted in detail. On the other hand, certain predictions have to be made still. In our example the genes are predicting that sweet taste in the mouth, and orgasm, are going to be 'good' in the sense that eating sugar and copulating are likely to be beneficial to gene survival. The possibilities of saccharine and masturbation are not anticipated according to this example; nor are the dangers of over-eating sugar in our environment where it exists in unnatural plenty.

Learning-strategies have been used in some chess-playing computer programs. These programs actually get better as they play against human opponents or against other computers. Although they are equipped with a repertoire of rules and tactics, they also have a small random tendency built into their decision procedure. They record past decisions, and whenever they win a game they slightly increase the weighting given to the tactics which preceded the victory, so that next time they are a little bit more likely to choose those same tactics again.

One of the most interesting methods of predicting the future is simulation. If a general wishes to know whether a particular military plan will be better than alternatives, he has a problem in prediction. There are unknown quantities in the weather, in the morale of his own troops, and in the possible countermeasures of the enemy. One way of discovering whether it is a good plan is to try it and see, but it is undesirable to use this test for all the tentative plans dreamed up, if only because the supply of young men prepared to die 'for their country' is exhaustible, and the supply of possible plans is very large. It is better to try the various plans out in dummy runs rather than in deadly earnest. This may take the form of full-scale exercises with 'Northland' fighting 'Southland' using blank ammunition, but even this is expensive in time and materials. Less wastefully, war games may be played, with tin soldiers and little toy tanks being shuffled around a large map.

Recently, computers have taken over large parts of the simulation function, not only in military strategy, but in all fields where prediction of the future is necessary, fields like economics, ecology, sociology, and many others. The technique works like this. A model of some aspect of the world is set up in the computer. This does not mean that if you unscrewed the lid you would see a little miniature dummy inside with the same shape as the object simulated. In the chess-playing computer there is no 'mental picture' inside the memory banks recognizable as a chess board with knights and pawns sitting on it. The chess board and its current position would be represented by lists of electronically coded numbers. To us a map is a miniature scale model of a part of the world, compressed into two dimensions. In a computer, a map would more probably be represented as a list of towns and other spots, each with two numbers—its latitude and longitude. But it does not matter how the computer actually holds its model of the world in its head, provided that it holds it in a form in which it can operate on it, manipulate it, do experiments with it, and report back to the human operators in terms which they can understand. Through the technique of simulation, model battles can be won or lost, simulated airliners fly or crash, economic policies lead to prosperity or to ruin. In each case the whole process goes on inside the computer in a tiny fraction of the time it would take in real life. Of course there are good models of the world and bad ones, and even the good ones are only approximations. No amount of simulation can predict exactly what will happen in reality, but a good simulation is enormously preferable to blind trial and error. Simulation could be called vicarious trial and error, a term unfortunately pre-empted long ago by rat psychologists.

If simulation is such a good idea, we might expect that survival machines would have discovered it first. After all, they invented many of the other techniques of human engineering long before we came on the scene: the focusing lens and the parabolic reflector, frequency analysis of sound waves, servo-control, sonar, buffer storage of incoming information, and countless others with long names, whose details don't matter. What about simulation? Well, when you yourself have a difficult decision to make involving unknown quantities in the future, you do go in for a form of simulation. You *imagine* what would happen if you did each of the alternatives open to you. You set up a model in your head, not of everything in the world, but of the restricted set of entities which you think may be relevant. You may see them vividly in your mind's eye, or you may see and manipulate stylized abstractions of them. In either case it is unlikely that somewhere laid out in your brain is an actual spatial model of the events you are imagining. But, just as in the computer, the details of how your brain represents its model of the world are less important than the fact that it is able to use it to predict possible events. Survival machines which can simulate the future are one jump ahead of survival machines who can only learn on the basis of overt trial and error. The trouble with overt trial is that it takes time and energy. The trouble with overt error is that it is often fatal. Simulation is both safer and faster.

The evolution of the capacity to simulate seems to have culminated in subjective consciousness. Why this should have happened is, to me, the most profound mystery facing modern biology. There is no reason to suppose that electronic computers are conscious when they simulate, although we have to admit that in the future they may become so. Perhaps consciousness arises when the brain's simulation of the world becomes so complete that it must include a model of itself. Obviously the limbs and body of a survival machine must constitute an important part of its simulated world; presumably for the same kind of reason, the simulation itself could be regarded as part of the world to be simulated. Another word for this might indeed be 'self-awareness', but I don't find this a fully satisfying explanation of the evolution of consciousness, and this is only partly because it involves an infinite regress—if there is a model of the model, why not a model of the model of the model . . .?

Whatever the philosophical problems raised by consciousness, for the purpose of this story it can be thought of as the culmination of an evolutionary trend towards the emancipation of survival machines as executive decision-takers from their ultimate masters, the genes. Not only are brains in charge of the day-to-day running of survival-machine affairs, they have also acquired the ability to predict the future and act accordingly. They even have the power to rebel against the dictates of the genes, for instance in refusing to have as many children as they are able to. But in this respect man is a very special case, as we shall see.

What has all this to do with altruism and selfishness? I am trying to build up the idea that animal behaviour, altruistic or selfish, is under the control of genes in only an indirect, but still very powerful, sense. By dictating the way survival machines and their nervous systems are built, genes exert ultimate power over behaviour. But the moment-to-moment decisions about what to do next are taken by the nervous system. Genes are the primary policy-makers; brains are the executives. But as brains became more highly developed, they took over more and more of the actual policy decisions, using tricks like learning and simulation in doing so. The logical conclusion to this trend, not yet reached in any species, would be for the genes to give the survival machine a single overall policy instruction: do whatever you think best to keep us alive.

Analogies with computers and with human decision-taking are all very well. But now we must come down to earth and remember that evolution in fact occurs step-by-step, through the differential survival of genes in the gene pool. Therefore, in order for a behaviour pattern—altruistic or selfish—to evolve, it is necessary that a gene 'for' that behaviour should survive in the gene pool more successfully than a rival gene or allele 'for' some different behaviour. A gene for altruistic behaviour means any gene which influences the development of nervous systems in such a way as to make them likely to behave altruistically. Is there any experimental evidence for the genetic inheritance of altruistic behaviour? No, but that is hardly surprising, since little work has been done on the genetics of any behaviour. Instead, let me tell you about one study of a behaviour pattern which does not happen to be obviously altruistic, but which is complex enough to be interesting. It serves as a model for how altruistic behaviour might be inherited.

Honey bees suffer from an infectious disease called foul brood. This attacks the grubs in their cells. Of the domestic breeds used by beekeepers, some are more at risk from foul brood than others, and it turns out that the difference between strains is, at least in some cases, a behavioural one. There are so-called hygienic strains which quickly stamp out epidemics by locating infected grubs, pulling them from their cells and throwing them out of the hive. The susceptible strains are susceptible because they do not practise this hygienic infanticide. The behaviour actually involved in hygiene is quite complicated. The workers have to locate the cell of each diseased grub, remove the wax cap from the cell, pull out the larva, drag it through the door of the hive, and throw it on the rubbish tip.

Doing genetic experiments with bees is quite a complicated business for various reasons. Worker bees themselves do not ordinarily reproduce, and so you have to cross a queen of one strain with a drone (= male) of the other, and then look at the behaviour of the daughter workers. This is what W. C. Rothenbuhler did. He found that all first-generation hybrid daughter hives were non-hygienic: the behaviour of their hygienic parent seemed to have been lost, although as things turned out the hygienic genes were still there but were recessive, like human genes for blue eyes. When Rothenbuhler 'back-crossed' first-generation hybrids with a pure hygienic strain (again of course using queens and drones), he obtained a most beautiful result. The daughter hives fell into three groups. One group showed perfect hygienic behaviour, a second showed no hygienic behaviour at all, and the third went half way. This last group uncapped the wax cells of diseased grubs, but they did not follow through and throw out the larvae. Rothenbuhler surmised that there might be two separate genes, one gene for uncapping, and one gene for throwing-out. Normal hygienic strains possess both genes, susceptible strains possess the alleles—rivals—of both genes instead. The hybrids who only went half way presumably possessed the uncapping gene (in double dose) but not the throwing-out gene. Rothenbuhler guessed that his experimental group of apparently totally non-hygienic bees might conceal a subgroup possessing the throwing-out gene, but unable to show it because they lacked the uncapping gene. He confirmed this most elegantly by removing caps himself. Sure enough, half of the apparently non-hygienic bees thereupon showed perfectly normal throwing-out behaviour.

This story illustrates a number of important points which came up in the previous chapter. It shows that it can be perfectly proper to speak of 'a gene for behaviour so-and-so' even if we haven't the faintest idea of the chemical chain of embryonic causes leading from gene to behaviour. The chain of causes could even turn out to involve learning. For example, it could be that the uncapping gene exerts its effect by giving bees a taste for infected wax. This means they will find the eating of the wax caps covering disease-victims rewarding, and will therefore tend to repeat it. Even if this is how the gene works, it is still truly a gene 'for uncapping' provided that, other things being equal, bees possessing the gene end up by uncapping, and bees not possessing the gene do not uncap.

Secondly it illustrates the fact that genes 'cooperate' in their effects on the behaviour of the communal survival machine. The throwing-out gene is useless unless it is accompanied by the uncapping gene and vice versa. Yet the genetic experiments show equally clearly that the two genes are in principle quite separable in their journey through the generations. As far as their useful work is concerned you can think of them as a single cooperating unit, but as replicating genes they are two free and independent agents.

For purposes of argument it will be necessary to speculate about genes 'for' doing all sorts of improbable things. If I speak, for example, of a hypothetical gene 'for saving companions from drowning', and you find such a concept incredible, remember the story of the hygienic bees. Recall that we are not talking about the gene as the sole antecedent cause of all the complex muscular contractions, sensory integrations, and even conscious decisions, which are involved in saving somebody from drowning. We are saying nothing about the question of whether learning, experience, or environmental influences enter into the development of the behaviour. All you have to concede is that it is possible for a single gene, other things being equal and lots of other essential genes and environmental factors being present, to make a body more likely to save somebody from drowning than its allele would. The difference between the two genes may turn out at bottom to be a slight difference in some simple quantitative variable. The details of the embryonic developmental process, interesting as they may be, are irrelevant to evolutionary considerations. Konrad Lorenz has put this point well.

The genes are master programmers, and they are programming

for their lives. They are judged according to the success of their programs in coping with all the hazards which life throws at their survival machines, and the judge is the ruthless judge of the court of survival. We shall come later to ways in which gene survival can be fostered by what appears to be altruistic behaviour. But the obvious first priorities of a survival machine, and of the brain that takes the decisions for it, are individual survival and reproduction. All the genes in the 'colony' would agree about these priorities. Animals therefore go to elaborate lengths to find and catch food; to avoid being caught and eaten themselves; to avoid disease and accident; to protect themselves from unfavourable climatic conditions; to find members of the opposite sex and persuade them to mate; and to confer on their children advantages similar to those they enjoy themselves. I shall not give examples—if you want one just look carefully at the next wild animal that you see. But I do want to mention one particular kind of behaviour because we shall need to refer to it again when we come to speak of altruism and selfishness. This is the behaviour that can be broadly labelled *communication*.

A survival machine may be said to have communicated with another one when it influences its behaviour or the state of its nervous system. This is not a definition I would like to have to defend for very long, but it is good enough for present purposes. By influence I mean direct causal influence. Examples of communication are numerous: song in birds, frogs, and crickets; tail-wagging and hackle-raising in dogs; 'grinning' in chimpanzees; human gestures and language. A great number of survival-machine actions promote their genes' welfare indirectly by influencing the behaviour of other survival machines. Animals go to great lengths to make this communication effective. The songs of birds enchant and mystify successive generations of men. I have already referred to the even more elaborate and mysterious song of the humpback whale, with its prodigious range, its frequencies spanning the whole of human hearing from subsonic rumblings to ultrasonic squeaks. Mole-crickets amplify their song to stentorian loudness by singing down in a burrow which they carefully dig in the shape of a double exponential horn, or megaphone. Bees dance in the dark to give other bees accurate information about the direction and distance of food, a feat of communication rivalled only by human language itself.

The traditional story of ethologists is that communication signals evolve for the mutual benefit of both sender and recipient. For instance, baby chicks influence their mother's behaviour by giving high piercing cheeps when they are lost or cold. This usually has the immediate effect of summoning the mother, who leads the chick back to the main clutch. This behaviour could be said to have evolved for mutual benefit, in the sense that natural selection has favoured babies who cheep when they are lost, and also mothers who respond appropriately to the cheeping.

If we wish to (it is not really necessary), we can regard signals such as the cheep call as having a meaning, or as carrying information: in this case 'I am lost.' The alarm call given by small birds, which I mentioned in Chapter 1, could be said to convey the information 'There is a hawk.' Animals who receive this information and act on it are benefited. Therefore the information can be said to be true. But do animals ever communicate false information; do they ever tell lies?

The notion of an animal telling a lie is open to misunderstanding, so I must try to forestall this. I remember attending a lecture given by Beatrice and Allen Gardner about their famous 'talking' chimpanzee Washoe (she uses American Sign Language, and her achievement is of great potential interest to students of language). There were some philosophers in the audience, and in the discussion after the lecture they were much exercised by the question of whether Washoe could tell a lie. I suspected that the Gardners thought there were more interesting things to talk about, and I agreed with them. In this book I am using words like 'deceive' and 'lie' in a much more straightforward sense than those philosophers. They were interested in conscious intention to deceive. I am talking simply about having an effect functionally equivalent to deception. If a bird used the 'There is a hawk' signal when there was no hawk, thereby frightening his colleagues away, leaving him to eat all their food, we might say he had told a lie. We would not mean he had deliberately intended consciously to deceive. All that is implied is that the liar gained food at the other birds' expense, and the reason the other birds flew away was that they reacted to the liar's cry in a way appropriate to the presence of a hawk.

Many edible insects, like the butterflies of the previous chapter, derive protection by mimicking the external appearance of other distasteful or stinging insects. We ourselves are often fooled into thinking that yellow and black striped hover-flies are wasps. Some bee-mimicking flies are even more perfect in their deception. Predators too tell lies. Angler fish wait patiently on the bottom of the sea, blending in with the background. The only conspicuous part is a wrigging worm-like piece of flesh on the end of a long 'fishing rod', projecting from the top of the head. When a small prey fish comes near, the angler will dance its worm-like bait in front of the little fish, and lure it down to the region of the angler's own concealed mouth. Suddenly it opens its jaws, and the little fish is sucked in and eaten. The angler is telling a lie, exploiting the little fish's tendency to approach wriggling worm-like objects. He is saying 'Here is a worm', and any little fish who 'believes' the lie is quickly eaten.

Some survival machines exploit the sexual desires of others. Bee orchids induce bees to copulate with their flowers, because of their strong resemblance to female bees. What the orchid has to gain from this deception is pollination, for a bee who is fooled by two orchids will incidentally carry pollen from one to the other. Fireflies (which are really beetles) attract their mates by flashing lights at them. Each species has its own particular dot-dash flashing pattern, which prevents confusion between species, and consequent harmful hybridization. Just as sailors look out for the flash patterns of particular lighthouses, so fireflies seek the coded flash patterns of their own species. Females of the genus *Photuris* have 'discovered' that they can lure males of the genus *Photinus* if they imitate the flashing code of a *Photinus* female. This they do, and when a *Photinus* male is fooled by the lie into approaching, he is summarily eaten by the *Photuris* female. Sirens and Lorelei spring to mind as analogies, but Cornishmen will prefer to think of the wreckers of the old days, who used lanterns to lure ships on to the rocks, and then plundered the cargoes which spilled out of the wrecks.

Whenever a system of communication evolves, there is always the danger that some will exploit the system for their own ends. Brought up as we have been on the 'good of the species' view of evolution, we naturally think first of liars and deceivers as belonging to different species: predators, prey, parasites, and so on. However, we must expect lies and deceit, and selfish exploitation of communication to arise whenever the interests of the genes of different individuals diverge. This will include individuals of the same species. As we shall see, we must even expect that children will deceive their parents, that husbands will cheat on wives, and that brother will lie to brother.

Even the belief that animal communication signals originally evolve to foster mutual benefit, and then afterwards become exploited by malevolent parties, is too simple. It may well be that all animal communication contains an element of deception right from the start, because all animal interactions involve at least some conflict of interest. The next chapter introduces a powerful way of thinking about conflicts of interest from an evolutionary point of view.

5. Aggression:
stability and the selfish machine

THIS chapter is mostly about the much-misunderstood topic of aggression. We shall continue to treat the individual as a selfish machine, programmed to do whatever is best for his genes as a whole. This is the language of convenience. At the end of the chapter we return to the language of single genes.

To a survival machine, another survival machine (which is not its own child or another close relative) is a part of its environment, like a rock or a river or a lump of food. It is something that gets in the way, or something that can be exploited. It differs from a rock or a river in one important respect: it is inclined to hit back. This is because it too is a machine which holds its immortal genes in trust for the future, and it too will stop at nothing to preserve them. Natural selection favours genes which control their survival machines in such a way that they make the best use of their environment. This includes making the best use of other survival machines, both of the same and of different species.

In some cases survival machines seem to impinge rather little on each others' lives. For instance moles and blackbirds do not eat each other, mate with each other, or compete with each other for living space. Even so, we must not treat them as completely insulated. They may compete for something, perhaps earthworms. This does not mean you will ever see a mole and a blackbird engaged in a tug of war over a worm; indeed a blackbird may never set eyes on a mole in its life. But if you wiped out the population of moles, the effect on blackbirds might be dramatic, although I could not hazard a guess as to what the details might be, nor by what tortuously indirect routes the influence might travel.

Survival machines of different species influence each other in a variety of ways. They may be predators or prey, parasites or hosts, competitors for some scarce resource. They may be exploited in special ways, as for instance when bees are used as pollen carriers by flowers.

Survival machines of the same species tend to impinge on each others' lives more directly. This is for many reasons. One is that half the population of one's own species may be potential mates, and potentially hard-working and exploitable parents to one's children. Another reason is that members of the same species, being very similar to each other, being machines for preserving genes in the same kind of place, with the same kind of way of life, are particularly direct competitors for all the resources necessary for life. To a blackbird, a mole may be a competitor, but it is not nearly so important a competitor as another blackbird. Moles and blackbirds may compete for worms, but blackbirds and blackbirds compete with each other for worms *and* for everything else. If they are members of the same sex, they may also compete for mating partners. For reasons which we shall see, it is usually the males who compete with each other for females. This means that a male might benefit his own genes if he does something detrimental to another male with whom he is competing.

The logical policy for a survival machine might therefore seem to be to murder its rivals, and then, preferably, to eat them. Although murder and cannibalism do occur in nature, they are not as common as a naïve interpretation of the selfish gene theory might predict. Indeed Konrad Lorenz, in *On Aggression*, stresses the restrained and gentlemanly nature of animal fighting. For him the notable thing about animal fights is that they are formal tournaments, played according to rules like those of boxing or fencing. Animals fight with gloved fists and blunted foils. Threat and bluff take the place of deadly earnest. Gestures of surrender are recognized by victors, who then refrain from dealing the killing blow or bite which our naïve theory might predict.

This interpretation of animal aggression as being restrained and formal can be disputed. In particular, it is certainly wrong to condemn poor old *Homo sapiens* as the only species to kill his own kind, the only inheritor of the mark of Cain, and similar melodramatic charges. Whether a naturalist stresses the violence or the restraint of animal aggression depends partly on the kinds of animals he is used to watching, and partly on his evolutionary preconceptions—Lorenz is, after all, a 'good of the species' man. Even if it has been exaggerated, the gloved fist view of animal fights seems to have at least some truth. Superficially this looks

like a form of altruism. The selfish gene theory must face up to
the difficult task of explaining it. Why is it that animals do not go
all out to kill rival members of their species at every possible
opportunity?

The general answer to this is that there are costs as well as
benefits resulting from outright pugnacity, and not only the
obvious costs in time and energy. For instance, suppose that B
and C are both my rivals, and I happen to meet B. It might seem
sensible for me as a selfish individual to try to kill him. But wait.
C is also my rival, and C is also B's rival. By killing B, I am
potentially doing a good turn to C by removing one of his rivals. I
might have done better to let B live, because he might then have
competed or fought with C, thereby benefiting me indirectly. The
moral of this simple hypothetical example is that there is no
obvious merit in indiscriminately trying to kill rivals. In a large
and complex system of rivalries, removing one rival from the
scene does not necessarily do any good: other rivals may be more
likely to benefit from his death than oneself. This is the kind of
hard lesson that has been learned by pest-control officers. You
have a serious agricultural pest, you discover a good way to exter-
minate it and you gleefully do so, only to find that another pest
benefits from the extermination even more than human agricul-
ture does, and you end up worse off than you were before.

On the other hand, it might seem a good plan to kill, or at least
fight with, certain particular rivals in a discriminating way. If B is
an elephant seal in possession of a large harem full of females, and
if I, another elephant seal, can acquire his harem by killing him, I
might be well advised to attempt to do so. But there are costs and
risks even in selectivity pugnacity. It is to B's advantage to fight
back, to defend his valuable property. If I start a fight, I am just
as likely to end up dead as he is. Perhaps even more so. He holds
a valuable resource, that is why I want to fight him. But why does
he hold it? Perhaps he won it in combat. He has probably beaten
off other challengers before me. He is probably a good fighter.
Even if I win the fight and gain the harem, I may be so badly
mauled in the process that I cannot enjoy the benefits. Also,
fighting uses up time and energy. These might be better concerved
for the time being. If I concentrate on feeding and on keeping out
of trouble for a time, I shall grow bigger and stronger. I'll fight

him for the harem in the end, but I may have a better chance of winning eventually if I wait, rather than rush in now.

This subjective soliloquy is just a way of pointing out that the decision whether or not to fight should ideally be preceded by a complex, if unconscious, 'cost–benefit' calculation. The potential benefits are not all stacked up on the side of fighting, although undoubtedly some of them are. Similarly, during a fight, each tactical decision over whether to escalate the fight or cool it has costs and benefits which could, in principle, be analysed. This has long been realized by ethologists in a vague sort of way, but it has taken J. Maynard Smith, not normally regarded as an ethologist, to express the idea forcefully and clearly. In collaboration with G. R. Price and G. A. Parker, he uses the branch of mathematics known as Game Theory. Their elegant ideas can be expressed in words without mathematical symbols, albeit at some cost in rigour.

The essential concept Maynard Smith introduces is that of the *evolutionarily stable strategy*, an idea which he traces back to W. D. Hamilton and R. H. MacArthur. A 'strategy' is a pre-programmed behavioural policy. An example of a strategy is: 'Attack opponent; if he flees pursue him; if he retaliates run away.' It is important to realize that we are not thinking of the strategy as being consciously worked out by the individual. Remember that we are picturing the animal as a robot survival machine with a pre-programmed computer controlling the muscles. To write the strategy out as a set of simple instructions in English is just a convenient way for us to think about it. By some unspecified mechanism, the animal behaves as if he were following these instructions.

An evolutionarily stable strategy or ESS is defined as a strategy which, if most members of a population adopt it, cannot be bettered by an alternative strategy. It is a subtle and important idea. Another way of putting it is to say that the best strategy for an individual depends on what the majority of the population are doing. Since the rest of the population consists of individuals, each one trying to maximize his *own* success, the only strategy that persists will be one which, once evolved, cannot be bettered by any deviant individual. Following a major environmental change there may be a brief period of evolutionary instability, perhaps even oscillation in the population. But once an ESS is achieved it will stay: selection will penalize deviation from it.

To apply this idea to aggression, consider one of Maynard Smith's simplest hypothetical cases. Suppose that there are only two sorts of fighting strategy in a population of a particular species, named *hawk* and *dove*. (The names refer to conventional human usage and have no connection with the habits of the birds from whom the names are derived: doves are in fact rather aggressive birds.) Any individual of our hypothetical population is classified as a hawk or a dove. Hawks always fight as hard and as unrestrainedly as they can, retreating only when seriously injured. Doves merely threaten in a dignified conventional way, never hurting anybody. If a hawk fights a dove the dove quickly runs away, and so does not get hurt. If a hawk fights a hawk they go on until one of them is seriously injured or dead. If a dove meets a dove nobody gets hurt; they go on posturing at each other for a long time until one of them tires or decides not to bother any more, and therefore backs down. For the time being, we assume that there is no way in which an individual can tell, in advance, whether a particular rival is a hawk or a dove. He only discovers this by fighting him, and he has no memory of past fights with particular individuals to guide him.

Now as a purely arbitrary convention we allot contestants 'points'. Say 50 points for a win, 0 for losing, −100 for being seriously injured, and −10 for wasting time over a long contest. These points can be thought of as being directly convertible into the currency of gene survival. An individual who scores high points, who has a high average 'pay-off', is an individual who leaves many genes behind him in the gene pool. Within broad limits the actual numerical values do not matter for the analysis, but they help us to think about the problem.

The important thing is that we are *not* interested in whether hawks will tend to beat doves when they fight them. We already know the answer to that: hawks will always win. We want to know whether either hawk or dove is an evolutionarily stable strategy. If one of them is an ESS and the other is not, we must expect that the one which is the ESS will evolve. It is theoretically possible for there to be two ESSs. This would be true if, whatever the majority strategy of the population happened to be, whether hawk or dove, the best strategy for any given individual was to follow suit. In this case the population would tend to stick at whichever one of its two stable states it happened to reach first. However, as we shall now see, neither of these two strategies, hawk or dove, would in fact be evolutionarily stable on its own, and we should therefore not expect either of them to evolve. To show this we must calculate average pay-offs.

Suppose we have a population consisting entirely of doves. Whenever they fight, nobody gets hurt. The contests consist of prolonged ritual tournaments, staring matches perhaps, which end only when one rival backs down. The winner then scores 50 points for gaining the resource in dispute, but he pays a penalty of −10 for wasting time over a long staring match, so scores 40 in all. The loser also is penalized −10 points for wasting time. On average, any one individual dove can expect to win half his contests and lose half. Therefore his average pay-off per contest is the average of +40 and −10, which is +15. Therefore, every individual dove in a population of doves seems to be doing quite nicely.

But now suppose a mutant hawk arises in the population. Since he is the only hawk around, every fight he has is against a dove. Hawks always beat doves, so he scores +50 every fight, and this is his average pay-off. He enjoys an enormous advantage over the doves, whose net pay-off is only +15. Hawk genes will rapidly spread through the population as a result. But now each hawk can no longer count on every rival he meets being a dove. To take an extreme example, if the hawk gene spread so successfully that the entire population came to consist of hawks, all fights would now be hawk fights. Things are now very different. When hawk meets hawk, one of them is seriously injured, scoring −100, while the winner scores +50. Each hawk in a population of hawks can expect to win half his fights and lose half his fights. His average expected pay-off per fight is therefore half-way between +50 and −100, which is −25. Now consider a single dove in a population of hawks. To be sure, he loses all his fights, but on the other hand he never gets hurt. His average pay-off is 0 in a population of hawks, whereas the average pay-off for a hawk in a population of hawks is −25. Dove genes will therefore tend to spread through the population.

The way I have told the story it looks as if there will be a continuous oscillation in the population. Hawk genes will sweep to ascendancy; then, as a consequence of the hawk majority, dove genes will gain an advantage and increase in numbers until once again hawk genes start to prosper, and so on. However, it need not be an oscillation like this. There is a stable ratio of hawks to doves. For the particular arbitrary points system we are using, the stable ratio, if you work it out, turns out to be $\frac{5}{12}$ doves to $\frac{7}{12}$ hawks. When this stable ratio is reached, the average pay-off for hawks is exactly equal to the average pay-off for doves. Therefore selection does not favour either one of them over the other. If the number of hawks in the population started to drift upwards so that the ratio was no longer $\frac{7}{12}$, doves would start to gain an extra advantage, and the ratio would swing back to the stable state. Just as we shall find the stable sex ratio to be 50 : 50, so the stable hawk to dove ratio in this hypothetical example is 7 : 5. In either case, if there are oscillations about the stable point, they need not be very large ones.

Superficially, this sounds a little like group selection, but it is really nothing of the kind. It sounds like group selection because it enables us to think of a population as having a stable equilibrium to which it tends to return when disturbed. But the ESS is a much more subtle concept than group selection. It has nothing to do with some groups being more successful than others. This can be nicely illustrated using the arbitrary points system of our hypothetical example. The average pay-off to an individual in a stable population consisting of $\frac{7}{12}$ hawks and $\frac{5}{12}$ doves, turns out to be $6\frac{1}{4}$. This is true whether the individual is a hawk or a dove. Now $6\frac{1}{4}$ is much less than the average pay-off for a dove in a population of doves (15). If *only* everybody would agree to be a dove, every single individual would benefit. By simple group selection, any group in which all individuals mutually agreed to be doves would be far more successful than a rival group sitting at the ESS ratio. (As a matter of fact, a conspiracy of nothing but doves is not quite the most successful possible group. In a group consisting of $\frac{1}{6}$ hawks and $\frac{5}{6}$ doves, the average pay-off per contest is $16\frac{2}{3}$. This is the most successful possible conspiracy, but for present purposes we can ignore it. A simpler all-dove conspiracy, with its average pay-off for each individual of 15, is far better for every single individual than the ESS would be.) Group selection theory would therefore predict a tendency to evolve towards an all-dove conspiracy, since a *group* which contained a $\frac{7}{12}$ proportion of hawks would be less successful. But the trouble with conspiracies, even those which are to everybody's advantage in the long run, is that they are open to abuse. It is true that everybody does better in an all-dove group than he would in an ESS group. But unfortunately, in conspiracies of doves, a single hawk does so extremely well that nothing could stop the evolution of hawks. The conspiracy is therefore bound to be broken by treachery from within. An ESS is stable, not because it is particularly good for the individuals participating in it, but simply because it is immune to treachery from within.

It is possible for humans to enter into pacts or conspiracies which are to every individual's advantage, even if these are not stable in the ESS sense. But this is only possible because every individual uses his *conscious* foresight, and is able to see that it is in his own long-term interests to obey the rules of the pact. Even in human pacts there is a constant danger that individuals will stand to gain so much in the *short term* by breaking the pact that the temptation to do so will be overwhelming. Perhaps the best example of this is price-fixing. It is in the long-term interests of all individual garage owners to standardize the price of petrol at some artificially high value. Price rings, based on conscious estimation of long-term best interests, can survive for quite long periods. Every so often, however, an individual gives in to the temptation to make a quick killing by cutting his prices. Immediately, his neighbours follow suit, and a wave of price cutting spreads over the country. Unfortunately for the rest of us, the conscious foresight of the garage owners then reasserts itself, and they enter into a new price-fixing pact. So, even in man, a species with the gift of conscious foresight, pacts or conspiracies based on long-term best interests teeter constantly on the brink of collapse due to treachery from within. In wild animals, controlled by the struggling genes, it is even more difficult to see ways in which group benefit or conspiracy strategies could possibly evolve. We must expect to find evolutionarily stable strategies everywhere.

In our hypothetical example we made the simple assumption

that any one individual was either a hawk or a dove. We ended up with an evolutionarily stable ratio of hawks to doves. In practice, what this means is that a stable ratio of hawk genes to dove genes would be achieved in the gene pool. The genetic technical term for this state is stable polymorphism. As far as the maths are concerned, an exactly equivalent ESS can be achieved without polymorphism as follows. If *every individual* is capable of behaving either like a hawk or like a dove in each particular contest, an ESS can be achieved in which all individuals have the same *probability* of behaving like a hawk, namely $\frac{7}{12}$ in our particular example. In practice this would mean that each individual enters each contest having made a random decision whether to behave on this occasion like a hawk or like a dove; random, but with a 7 : 5 bias in favour of hawk. It is very important that the decisions, although biased towards hawk, should be random in the sense that a rival has no way of guessing how his opponent is going to behave in any particular contest. It is no good, for instance, playing hawk seven fights in a row, then dove five fights in a row and so on. If any individual adopted such a simple sequence, his rivals would quickly catch on and take advantage. The way to take advantage of a simple sequence strategist is to play hawk against him only when you know he is going to play dove.

The hawk and dove story is, of course, naïvely simple. It is a 'model', something which does not really happen in nature, but which helps us understand things which do happen in nature. Models can be very simple, like this one, and still be useful for understanding a point, or getting an idea. Simple models can be elaborated and gradually made more complex. If all goes well, as they get more complex they come to resemble the real world more. One way in which we can begin to develop the hawk and dove model is to introduce some more strategies. Hawk and dove are not the only possibilities. A more complex strategy which Maynard Smith and Price introduced is called *Retaliator*.

A retaliator plays like a dove at the beginning of every fight. That is, he does not mount an all-out savage attack like a hawk, but has a conventional threatening match. If his opponent attacks him, however, he retaliates. In other words, a retaliator behaves like a hawk when he is attacked by a hawk, and like a dove when he meets a dove. When he meets another retaliator he plays like a dove. A retaliator is a *conditional strategist*. His behaviour depends on the behaviour of his opponent.

Another conditional strategist is called *Bully*. A bully goes around behaving like a hawk until somebody hits back. Then he immediately runs away. Yet another conditional strategist is *Prober–retaliator*. A prober–retaliator is basically like a retaliator, but he occasionally tries a brief experimental escalation of the contest. He persists in this hawk-like behaviour if his opponent does not fight back. If, on the other hand, his opponent does fight back he reverts to conventional threatening like a dove. If he is attacked, he retaliates just like an ordinary retaliator.

If all the five strategies I have mentioned are turned loose upon one another in a computer simulation, only one of them, retaliator, emerges as evolutionarily stable. Prober–retaliator is nearly stable. Dove is not stable, because a population of doves would be invaded by hawks and bullies. Hawk is not stable, because a population of hawks would be invaded by doves and bullies. Bully is not stable, because a population of bullies would be invaded by hawks. In a population of retaliators, no other strategy would invade, since there is no other strategy that does better than retaliator itself. However, dove does equally well in a population of retaliators. This means that, other things being equal, the numbers of doves could slowly drift upwards. Now if the numbers of doves drifted up to any significant extent, prober–retaliators (and, incidentally, hawks and bullies) would start to have an advantage, since they do better against doves than retaliators do. Prober–retaliator itself, unlike hawk and bully, is almost an ESS, in the sense that, in a population of prober–retaliators, only one other strategy, retaliator, does better, and then only slightly. We might expect, therefore, that a mixture of retaliators and prober–retaliators would tend to predominate, with perhaps even a gentle oscillation between the two, in association with an oscillation in the size of a small dove minority. Once again, we don't have to think in terms of a polymorphism in which every individual always plays one strategy or another. Each individual could play a complex mixture between retaliator, prober–retaliator, and dove.

This theoretical conclusion is not far from what actually

happens in most wild animals. We have in a sense explained the
'gloved fist' aspect of animal aggression. Of course the details
depend on the exact numbers of 'points' awarded for winning,
being injured, wasting time, and so on. In elephant seals the prize
for winning may be near-monopoly rights over a large harem of
females. The pay-off for winning must therefore be rated as very
high. Small wonder that fights are vicious and the probability of
serious injury is also high. The cost of wasting time should
presumably be regarded as small in comparison with the cost of
being injured and the benefit of winning. For a small bird in a
cold climate, on the other hand, the cost of wasting time may be
paramount. A great tit when feeding nestlings needs to catch an
average of one prey per thirty seconds. Every second of daylight
is precious. Even the comparatively short time wasted in a hawk/
hawk fight should perhaps be regarded as more serious than the
risk of injury to such a bird. Unfortunately, we know too little at
present to assign realistic numbers to the costs and benefits of
various outcomes in nature. We must be careful not to draw
conclusions which result simply from our own arbitrary choice of
numbers. The general conclusions which are important are that
ESSs will tend to evolve, that an ESS is not the same as the
optimum which could be achieved by a group conspiracy, and
that common sense can be misleading.

Another kind of war game which Maynard Smith has considered is the 'war of attrition'. This can be thought of as arising
in a species which never engages in dangerous combat, perhaps a
well-armoured species in which injury is very unlikely. All
disputes in this species are settled by conventional posturing. A
contest always ends in one rival or the other backing down. To
win, all you have to do is stand your ground and glare at the
opponent until he finally turns tail. Obviously no animal can
afford to spend infinite time threatening; there are important
things to be done elsewhere. The resource he is competing for
may be valuable, but it is not infinitely valuable. It is only worth
so much time and, as at an auction sale, each individual is
prepared to spend only so much on it. Time is the currency of
this two-bidder auction.

Suppose all such individuals worked out in advance exactly
how much time they thought a particular kind of resource, say a

female, was worth. A mutant individual who was prepared to go on just a little bit longer would always win. So the strategy of maintaining a fixed bidding limit is unstable. Even if the value of the resource can be very finely estimated, and all individuals bid exactly the right value, the strategy is unstable. Any two individuals bidding according to this maximum strategy would give up at exactly the same instant, and neither would get the resource! It would then pay an individual to give up right at the start rather than waste any time in contests at all. The important difference between the war of attrition and a real auction sale is, after all, that in the war of attrition *both* contestants pay the price but only one of them gets the goods. In a population of maximum bidders, therefore, a strategy of giving up at the beginning would be successful and would spread through the population. As a consequence of this some benefit would start to accrue to individuals who did not give up immediately, but waited for a few seconds before giving up. This strategy would pay when played against the immediate retreaters who now predominate in the population. Selection would then favour a progressive extension of the giving-up time until it once more approached the maximum allowed by the true economic worth of the resource under dispute.

Once again, by using words, we have talked ourselves into picturing an oscillation in a population. Once again, mathematical analysis shows that this is not necessary. There is an evolutionarily stable strategy, which can be expressed as a mathematical formula, but in words what it amounts to is this. Each individual goes on for an *unpredictable* time. Unpredictable on any particular occasion, that is, but averaging the true value of the resource. For example, suppose the resource is really worth five minutes of display. At the ESS, any particular individual may go on for more than five minutes or he may go on for less than five minutes, or he may even go on for exactly five minutes. The important thing is that his opponent has no way of knowing how long he is prepared to persist on this particular occasion.

Obviously, it is vitally important in the war of attrition that individuals should give no inkling of when they are going to give up. Anybody who betrayed, by the merest flicker of a whisker, that he was beginning to think of throwing in the sponge, would be at an instant disadvantage. If, say, whisker-flickering happened to be a reliable sign that retreat would follow within one minute, there would be a very simple winning strategy: 'If your opponent's whiskers flicker, wait one more minute, regardless of what your own previous plans for giving up might have been. If your opponent's whiskers have not yet flickered, and you are within one minute of the time when you intend to give up anyway, give up immediately and don't waste any more time. Never flicker your own whiskers.' So natural selection would quickly penalize whisker-flickering and any analogous betrayals of future behaviour. The poker face would evolve.

Why the poker face rather than out-and-out lies? Once again, because lying is not stable. Suppose it happened to be the case that the majority of individuals raised their hackles only when they were truly intending to go on for a very long time in the war of attrition. The obvious counterploy would evolve: individuals would give up immediately when an opponent raised his hackles. But now, liars might start to evolve. Individuals who really had no intention of going on for a long time would raise their hackles on every occasion, and reap the benefits of easy and quick victory. So liar genes would spread. When liars became the majority, selection would now favour individuals who called their bluff. Therefore liars would decrease in numbers again. In the war of attrition, telling lies is no more evolutionarily stable than telling the truth. The poker face is evolutionarily stable. Surrender, when it finally comes, will be sudden and unpredictable.

So far we have considered only what Maynard Smith calls 'symmetric' contests. This means we have assumed that the contestants are identical in all respects except their fighting strategy. Hawks and doves are assumed to be equally strong, to be equally well endowed with weapons and with armour, and to have an equal amount to gain from winning. This is a convenient assumption to make for a model, but it is not very realistic. Parker and Maynard Smith went on to consider asymmetric contests. For example, if individuals vary in size and fighting ability, and each individual is capable of gauging a rival's size in comparison to his own, does this affect the ESS which emerges? It most certainly does.

There seem to be three main sorts of asymmetry. The first we have just met: individuals may differ in their size or fighting equipment. Secondly, individuals may differ in how much they have to gain from winning. For instance an old male, who has not long to live anyway, might have less to lose if he is injured than a young male with the bulk of his reproductive life ahead of him.

Thirdly, it is a strange consequence of the theory that a purely arbitrary, apparently irrelevant, asymmetry can give rise to an ESS, since it can be used to settle contests quickly. For instance it will usually be the case that one contestant happens to arrive at the location of the contest earlier than the other. Call them 'resident' and 'intruder' respectively. For the sake of argument, I am assuming that there is no general advantage attached to being a resident or an intruder. As we shall see, there are practical reasons why this assumption may not be true, but that is not the point. The point is that even if there were no general reason to suppose that residents have an advantage over intruders, an ESS depending on the asymmetry itself would be likely to evolve. A simple analogy is to humans who settle a dispute quickly and without fuss by tossing a coin.

The conditional strategy: 'If you are the resident, attack; if you are the intruder, retreat', could be an ESS. Since the asymmetry is assumed to be arbitrary, the opposite strategy: 'If resident, retreat; if intruder, attack' could also be stable. Which of the two ESSs is adopted in a particular population would depend on which one happens to reach a majority first. Once a majority of individuals is playing one of these two conditional strategies, deviants from it are penalized. Hence, by definition, it is an ESS.

For instance, suppose all individuals are playing 'resident wins, intruder runs away'. This means they will win half their fights and lose half their fights. They will never be injured and they will never waste time, since all disputes are instantly settled by arbitrary convention. Now consider a new mutant rebel. Suppose he plays a pure hawk strategy, always attacking and never retreating. He will win when his opponent is an intruder. When his opponent is a resident he will run a grave risk of injury. On average he will have a lower pay-off than individuals playing according to the arbitrary rules of the ESS. A rebel who tries the reverse convention 'if resident run away, if intruder attack', will do even worse. Not only will he frequently be injured, he will also seldom win a contest. Suppose, though, that by some chance events individuals playing this reverse convention managed to become the majority. In this case their strategy would then become the stable norm, and deviation from *it* would be penalized. Conceivably, if we watched a population for many generations we would see a series of occasional flips from one stable state to the other.

However, in real life, truly arbitrary asymmetries probably do not exist. For instance, residents probably tend to have a practical advantage over intruders. They have better knowledge of local terrain. An intruder is perhaps more likely to be out of breath because he moved into the battle area, whereas the resident was there all the time. There is a more abstract reason why, of the two stable states, the 'resident wins, intruder retreats' one is the more probable in nature. This is that the reverse strategy, 'intruder wins, resident retreats' has an inherent tendency to self-destruction—it is what Maynard Smith would call a paradoxical strategy. In any population sitting at this paradoxical ESS, individuals would always be striving never to be caught as residents: they would always be trying to be the intruder in any encounter. They could only achieve this by ceaseless, and otherwise pointless, moving around! Quite apart from the costs in time and energy that would be incurred, this evolutionary trend would, of itself, tend to lead to the category 'resident' ceasing to exist. In a population sitting at the other stable state, 'resident wins, intruder retreats', natural selection would favour individuals who strove to be residents. For each individual, this would mean holding on to a particular piece of ground, leaving it as little as possible, and appearing to 'defend' it. As is now well known, such behaviour is commonly observed in nature, and goes by the name of 'territorial defence'.

The neatest demonstration I know of this form of behavioural asymmetry was provided by the great ethologist Niko Tinbergen, in an experiment of characteristically ingenious simplicity. He had a fish-tank containing two male sticklebacks. The males had each built nests, at opposite ends of the tank, and each 'defended' the territory around his own nest. Tinbergen placed each of the two males in a large glass test-tube, and he held the two tubes next to each other and watched the males trying to fight each other through the glass. Now comes the interesting result. When he moved the two tubes into the vicinity of male A's nest, male A assumed an attacking posture, and male B attempted to retreat. But when he moved the two tubes into male B's territory, the tables were turned. By simply moving the two tubes from one end of the tank to the other, Tinbergen was able to dictate which male attacked and which retreated. Both males were evidently playing the simple conditional strategy: 'if resident, attack; if intruder, retreat.'

Biologists often ask what the biological 'advantages' of territorial behaviour are. Numerous suggestions have been made, some of which will be mentioned later. But we can now see that the very question may be superfluous. Territorial 'defence' may simply be an ESS which arises because of the asymmetry in time of arrival which usually characterizes the relationship between two individuals and a patch of ground.

Presumably the most important kind of non-arbitrary asymmetry is in size and general fighting ability. Large size is not necessarily always the most important quality needed to win fights, but it is probably one of them. If the larger of two fighters always wins, and if each individual knows for certain whether he is larger or smaller than his opponent, only one strategy makes any sense: 'If your opponent is larger than you, run away. Pick fights with people smaller than you are.' Things are a bit more complicated if the importance of size is less certain. If large size confers only a slight advantage, the strategy I have just mentioned is still stable. But if the risk of injury is serious there may also be a second, 'paradoxical strategy'. This is: 'Pick fights with people larger than you are and run away from people smaller than you are'! It is obvious why this is called paradoxical. It seems completely counter to common sense. The reason it can be stable is this. In a population consisting entirely of paradoxical strategists, nobody ever gets hurt. This is because in every contest one of the participants, the larger, always runs away. A mutant of average size who plays the 'sensible' strategy of picking on smaller opponents is involved in a seriously escalated fight with half the people he meets. This is because, if he meets somebody smaller than him, he attacks; the smaller individual fights back fiercely, because he is playing paradoxical; although the sensible strategist is more likely to win than the paradoxical one, he still runs a substantial risk of losing and of being seriously injured. Since the majority of the population are paradoxical, a sensible strategist is more likely to be injured than any single paradoxical strategist.

Even though a paradoxical strategy can be stable, it is probably only of academic interest. Paradoxical fighters will only have a higher average pay-off if they very heavily out-number sensible ones. It is hard to imagine how this state of affairs could ever arise in the first place. Even if it did, the ratio of sensibles to paradoxicals in the population only has to drift a little way towards the sensible side before reaching the 'zone of attraction' of the other ESS, the sensible one. The zone of attraction is the set of population ratios at which, in this case, sensible strategists have the advantage: once a population reaches this zone, it will be sucked inevitably towards the sensible stable point. It would be exciting to find an example of a paradoxical ESS in nature, but I doubt if we can really hope to do so. (I spoke too soon. After I had written this last sentence, Professor Maynard Smith called my attention to the following description of the behaviour of the Mexican social spider *Oecobius civitas*, by J. W. Burgess: 'If a spider is disturbed and driven out of its retreat, it darts across the rock and, in the absence of a vacant crevice to hide in, may seek refuge in the hiding place of another spider of the same species. If the other spider is in residence when the intruder enters, it does not attack but darts out and seeks a new refuge of its own. Thus once the first spider is disturbed the process of sequential displacement from web to web may continue for several seconds, often causing a majority of the spiders in the aggregation to shift from their home refuge to an alien one' (Social Spiders, *Scientific American*, March 1976). This is paradoxical in the sense of page 85.)

What if individuals retain some memory of the outcome of past fights? This depends on whether the memory is specific or general. Crickets have a general memory of what happened in past fights. A cricket which has recently won a large number of fights becomes more hawkish. A cricket which has recently had a losing streak becomes more dovish. This was neatly shown by R. D.

Alexander. He used a model cricket to beat up real crickets. After this treatment the real crickets became more likely to lose fights against other real crickets. Each cricket can be thought of as constantly updating his own estimate of his fighting ability, relative to that of an average individual in his population. If animals such as crickets, who work with a general memory of past fights, are kept together in a closed group for a time, a kind of dominance hierarchy is likely to develop. An observer can rank the individuals in order. Individuals lower in the order tend to give in to individuals higher in the order. There is no need to suppose that the individuals recognize each other. All that happens is that individuals who are accustomed to winning become even more likely to win, while individuals who are accustomed to losing become steadily more likely to lose. Even if the individuals started by winning or losing entirely at random, they would tend to sort themselves out into a rank order. This incidentally has the effect that the number of serious fights in the group gradually dies down.

I have to use the phrase 'kind of dominance hierarchy', because many people reserve the term dominance hierarchy for cases in which individual recognition is involved. In these cases, memory of past fights is specific rather than general. Crickets do not recognize each other as individuals, but hens and monkeys do. If you are a monkey, a monkey who has beaten you in the past is likely to beat you in the future. The best strategy for an individual is to be relatively dovish towards an individual who has previously beaten him. If a batch of hens who have never met before are introduced to each other, there is usually a great deal of fighting. After a time the fighting dies down. Not for the same reason as in the crickets, though. In the case of the hens it is because each individual 'learns her place' relative to each other individual. This is incidentally good for the group as a whole. As an indicator of this it has been noticed that in established groups of hens, where fierce fighting is rare, egg production is higher than in groups of hens whose membership is continually being changed, and in which fights are consequently more frequent. Biologists often speak of the biological advantage or 'function' of dominance hierarchies as being to reduce overt aggression in the group. However, this is the wrong way to put it. A dominance hierarchy *per se* cannot be said to have a 'function' in the evolutionary sense, since it is a property of a group, not of an individual. The individual behaviour patterns which manifest themselves in the form of dominance hierarchies when viewed at the group level may be said to have functions. It is, however, even better to abandon the word 'function' altogether, and to think about the matter in terms of ESSs in asymmetric contests where there is individual recognition and memory.

We have been thinking of contests between members of the same species. What about inter-specific contests? As we saw earlier, members of different species are less direct competitors than members of the same species. For this reason we should expect fewer disputes between them over resources, and our expectation is borne out. For instance, robins defend territories against other robins, but not against great tits. One can draw a map of the territories of different individual robins in a wood and one can superimpose a map of the territories of individual great tits. The territories of the two species overlap in an entirely indiscriminate way. They might as well be on different planets.

But there are other ways in which the interests of individuals from different species conflict very sharply. For instance a lion wants to eat an antelope's body, but the antelope has very different plans for its body. This is not normally regarded as competition for a resource, but logically it is hard to see why not. The resource in question is meat. The lion genes 'want' the meat as food for their survival machine. The antelope genes want the meat as working muscle and organs for their survival machine. These two uses for the meat are mutually incompatible, therefore there is conflict of interest.

Members of one's own species are made of meat too. Why is cannibalism relatively rare? As we saw in the case of black-headed gulls, adults do sometimes eat the young of their own species. Yet adult carnivores are never to be seen actively pursuing other adults of their own species with a view to eating them. Why not? We are still so used to thinking in terms of the 'good of the species' view of evolution that we often forget to ask perfectly reasonable questions like: 'Why don't lions hunt other lions?' Another good question of a type which is seldom asked is: 'Why do antelopes run away from lions instead of hitting back?'

The reason lions do not hunt lions is that it would not be an ESS for them to do so. A cannibal strategy would be unstable for the same reason as the hawk strategy in the earlier example. There is too much danger of retaliation. This is less likely to be true in contests between members of different species, which is why so many prey animals run away instead of retaliating. It probably stems originally from the fact that in an interaction between two animals of different species there is a built-in asymmetry which is greater than that between members of the same species. Whenever there is strong asymmetry in a contest, ESSs are likely to be conditional strategies dependent on the asymmetry. Strategies analogous to 'if smaller, run away; if larger, attack' are very likely to evolve in contests between members of different species because there are so many available asymmetries. Lions and antelopes have reached a kind of stability by evolutionary divergence, which has accentuated the original asymmetry of the contest in an ever-increasing fashion. They have become highly proficient in the arts of, respectively, chasing, and running away. A mutant antelope which adopted a 'stand and fight' strategy against lions would be less successful than rival antelopes disappearing over the horizon.

I have a hunch that we may come to look back on the invention of the ESS concept as one of the most important advances in evolutionary theory since Darwin. It is applicable wherever we find conflict of interest, and that means almost everywhere. Students of animal behaviour have got into the habit of talking about something called 'social organization'. Too often the social organization of a species is treated as an entity in its own right, with its own biological 'advantage'. An example I have already given is that of the 'dominance hierarchy'. I believe it is possible to discern hidden group-selectionist assumptions lying behind a large number of the statements which biologists make about social organization. Maynard Smith's concept of the ESS will enable us, for the first time, to see clearly how a collection of independent selfish entities can come to resemble a single organized whole. I think this will be true not only of social organizations within species, but also of 'ecosystems' and 'communities' consisting of many species. In the long term, I expect the ESS concept to revolutionize the science of ecology.

We can also apply it to a matter which was deferred from Chapter 3, arising from the analogy of oarsmen in a boat (representing genes in a body) needing a good team spirit. Genes are selected, not as 'good' in isolation, but as good at working against the background of the other genes in the gene pool. A good gene must be compatible with, and complementary to, the other genes with whom it has to share a long succession of bodies. A gene for plant-grinding teeth is a good gene in the gene pool of a herbivorous species, but a bad gene in the gene pool of a carnivorous species.

It is possible to imagine a compatible combination of genes as being selected together *as a unit*. In the case of the butterfly mimicry example of Chapter 3, this seems to be exactly what happened. But the power of the ESS concept is that it can now enable us to see how the same kind of result could be achieved by selection purely at the level of the independent gene. The genes do not have to be linked on the same chromosome.

The rowing analogy is really not up to explaining this idea. The nearest we can come to it is this. Suppose it is important in a really successful crew that the rowers should coordinate their activities by means of speech. Suppose further that, in the pool of oarsmen at the coach's disposal, some speak only English and some speak only German. The English are not consistently better or worse rowers than the Germans. But because of the importance of communication, a mixed crew will tend to win fewer races than either a pure English crew or a pure German crew.

The coach does not realize this. All he does is shuffle his men around, giving credit points to individuals in winning boats, marking down individuals in losing boats. Now if the pool available to him just happens to be dominated by Englishmen it follows that any German who gets into a boat is likely to cause it to lose, because communication breaks down. Conversely, if the pool happened to be dominated by Germans, an Englishman would tend to cause any boat in which he found himself to lose. What will emerge as the overall best crew will be one of the two stable states—pure English or pure German, but not mixed. Superficially it looks as though the coach is selecting whole language groups *as units*. This is not what he is doing. He is selecting individual oarsmen for their apparent ability to win races. It so happens that the tendency for an individual to win races depends on which other individuals are present in the pool of candidates. Minority candidates are automatically penalized, not because they are bad rowers, but simply because they are minority candidates. Similarly, the fact that genes are selected for mutual compatibility does not necessarily mean we *have* to think of groups of genes as being selected as units, as they were in the case of the butterflies. Selection at the low level of the single gene can give the impression of selection at some higher level.

In this example, selection favours simple conformity. More interestingly, genes may be selected because they complement each other. In terms of the analogy, suppose an ideally balanced crew would consist of four right-handers and four left-handers. Once again assume that the coach, unaware of this fact, selects blindly on 'merit'. Now if the pool of candidates happens to be dominated by right-handers, any individual left-hander will tend to be at an advantage: he is likely to cause any boat in which he finds himself to win, and he will therefore appear to be a good oarsman. Conversely, in a pool dominated by left-handers, a right-hander would have an advantage. This is similar to the case of a hawk doing well in a population of doves, and a dove doing well in a population of hawks. The difference is that there we were talking about interactions between individual bodies— selfish machines— whereas here we are talking, by analogy, about interactions between genes within bodies.

The coach's blind selection of 'good' oarsmen will lead in the end to an ideal crew consisting of four left-handers and four right-handers. It will look as though he selected them all together as a complete, balanced unit. I find it more parsimonious to think of him as selecting at a lower level, the level of the independent candidates. The evolutionarily stable state ('strategy' is misleading in this context) of four left-handers and four right-handers will emerge simply as a consequence of low-level selection on the basis of apparent merit.

The gene pool is the long-term environment of the gene. 'Good' genes are blindly selected as those which survive in the gene pool. This is not a theory; it is not even an observed fact: it is a tautology. The interesting question is what makes a gene good. As a first approximation I said that what makes a gene good is the ability to build efficient survival machines—bodies. We must now amend that statement. The gene pool will become an *evolutionarily stable set* of genes, defined as a gene pool which cannot be invaded by any new gene. Most new genes which arise, either by mutation or reassortment or immigration, are quickly penalized by natural selection: the evolutionarily stable set is restored. Occasionally a new gene does succeed in invading the set: it succeeds in spreading through the gene pool. There is a transitional period of instability, terminating in a new evolutionarily stable set—a little bit of evolution has occurred. By analogy with the aggression strategies, a population might have more than one alternative stable point, and it might occasionally flip from one to another. Progressive evolution may be not so much a steady upward climb as a series of discrete steps from stable plateau to stable plateau. It may look as though the population as a whole is behaving like a single self-regulating unit. But this illusion is produced by selection going on at the level of the single gene. Genes are selected on 'merit'. But merit is judged on the basis of performance against the background of the evolutionarily stable set which is the current gene pool.

By focusing on aggressive interactions between whole individuals, Maynard Smith was able to make things very clear. It is easy to think of stable ratios of hawk bodies and dove bodies, because bodies are large things which we can see. But such interactions between genes sitting in *different* bodies are only the tip of the iceberg. The vast majority of significant interactions between genes in the evolutionarily stable set—the gene pool—go on *within* individual bodies. These interactions are difficult to see, for they take place within cells, notably the cells of developing embryos. Well-integrated bodies exist because they are the product of an evolutionarily stable set of selfish genes.

But I must return to the level of interactions between whole animals which is the main subject of this book. For understanding aggression it was convenient to treat individual animals as independent selfish machines. This model breaks down when the individuals concerned are close relatives—brothers and sisters, cousins, parents and children. This is because relatives share a substantial proportion of their genes. Each selfish gene therefore has its loyalties divided between different bodies. This is explained in the next chapter.

6. Genesmanship

WHAT is the selfish gene? It is not just one single physical bit of DNA. Just as in the primeval soup, it is *all replicas* of a particular bit of DNA, distributed throughout the world. If we allow ourselves the licence of talking about genes as if they had conscious aims, always reassuring ourselves that we could translate our sloppy language back into respectable terms if we wanted to, we can ask the question, what is a single selfish gene trying to do? It is trying to get more numerous in the gene pool. Basically it does this by helping to program the bodies in which it finds itself to survive and to reproduce. But now we are emphasizing that 'it' is a distributed agency, existing in many different individuals at once. The key point of this chapter is that a gene might be able to assist *replicas* of itself which are sitting in other bodies. If so, this would appear as individual altruism but it would be brought about by gene selfishness.

Consider the gene for being an albino in man. In fact several genes exist which can give rise to albinism, but I am talking about just one of them. It is recessive; that is, it has to be present in double dose in order for the person to be an albino. This is true of about 1 in 20 000 of us. But it is also present, in single dose, in about 1 in 70 of us, and these individuals are not albinos. Since it is distributed in many individuals, a gene such as the albino gene could, in theory, assist its own survival in the gene pool by programming its bodies to behave altruistically towards other albino bodies, since these are known to contain the same gene. The albino gene should be quite happy if some of the bodies which it inhabits die, provided that in doing so they help other bodies containing the same gene to survive. If the albino gene could make one of its bodies save the lives of ten albino bodies, then even the death of the altruist is amply compensated by the increased numbers of albino genes in the gene pool.

Should we then expect albinos to be especially nice to each

other? Actually the answer is probably no. In order to see why not, we must temporarily abandon our metaphor of the gene as a conscious agent, because in this context it becomes positively misleading. We must translate back into respectable, if more longwinded terms. Albino genes do not really 'want' to survive or to help other albino genes. But if the albino gene just happened to cause its bodies to behave altruistically towards other albinos, then automatically, willy-nilly, it would tend to become more numerous in the gene pool as a result. But, in order for this to happen, the gene would have to have two independent effects on bodies. Not only must it confer its usual effect of a very pale complexion. It must also confer a tendency to be selectively altruistic towards individuals with a very pale complexion. Such a double-effect gene could, if it existed, be very successful in the population.

Now it is true that genes do have multiple effects, as I emphasized in Chapter 3. It is theoretically possible that a gene could arise which conferred an externally visible 'label', say a pale skin, or a green beard, or anything conspicuous, and also a tendency to be specially nice to bearers of that conspicuous label. It is possible, but not particularly likely. Green beardedness is just as likely to be linked to a tendency to develop ingrowing toenails or any other trait, and a fondness for green beards is just as likely to go together with an inability to smell freesias. It is not very probable that one and the same gene would produce both the right label and the right sort of altruism. Nevertheless, what may be called the Green Beard Altruism Effect is a theoretical possibility.

An arbitrary label like a green beard is just one way in which a gene might 'recognize' copies of itself in other individuals. Are there any other ways? A particularly direct possible way is the following. The possessor of an altruistic gene might be recognized simply by the fact that he does altruistic acts. A gene could prosper in the gene pool if it 'said' the equivalent of: 'Body, if A is drowning as a result of trying to save someone else from drowning, jump in and rescue A.' The reason such a gene could do well is that there is a greater than average chance that A contains the same life-saving altruistic gene. The fact that A is seen to be trying to rescue somebody else is a label, equivalent to a green beard. It is less arbitrary than a green beard, but it still seems rather implausible. Are there any plausible ways in which genes might 'recognize' their copies in other individuals?

The answer is yes. It is easy to show that *close relatives*—kin— have a greater than average chance of sharing genes. It has long been clear that this must be why altruism by parents towards their young is so common. What R. A. Fisher, J. B. S. Haldane, and especially W. D. Hamilton realized, was that the same applies to other close relations—brothers and sisters, nephews and nieces, close cousins. If an individual dies in order to save ten close relatives, one copy of the kin-altruism gene may be lost, but a larger number of copies of the same gene is saved.

'A larger number' is a bit vague. So is 'close relatives'. We can do better than that, as Hamilton showed. His two papers of 1964 are among the most important contributions to social ethology ever written, and I have never been able to understand why they have been so neglected by ethologists (his name does not even appear in the index of two major text-books of ethology, both published in 1970). Fortunately there are recent signs of a revival of interest in his ideas. Hamilton's papers are rather mathematical, but it is easy to grasp the basic principles intuitively, without rigorous mathematics, though at the cost of some over-simplification. The thing we want to calculate is the probability, or odds, that two individuals, say two sisters, share a particular gene.

For simplicity I shall assume that we are talking about genes which are rare in the gene pool as a whole. Most people share 'the gene for not being an albino', whether they are related to each other or not. The reason this gene is so common is that in nature albinos are less likely to survive than non-albinos because, for example, the sun dazzles them and makes them relatively unlikely to see an approaching predator. We are not concerned with explaining the prevalence in the gene pool of such obviously 'good' genes as the gene for not being an albino. We are interested in explaining the success of genes specifically as a result of their altruism. We can therefore assume that, at least in the early stages of this process of evolution, these genes are rare. Now the important point is that even a gene which is rare in the population as a whole is common within a family. I contain a number of genes which are rare in the population as a whole, and you also contain genes which are rare in the population as a whole. The chance that we both contain the same rare genes is very small indeed. But the chances are good that my sister contains a particularly rare gene that I contain, and the chances are equally good that your sister contains a rare gene in common with you. The odds are in this case exactly 50 per cent, and it is easy to explain why.

Suppose you contain one copy of the gene G. You must have received it either from your father or from your mother (for convenience we can neglect various infrequent possibilities—that G is a new mutation, that both your parents had it, or that either of your parents had two copies of it). Suppose it was your father who gave you the gene. Then every one of his ordinary body cells contained one copy of G. Now you will remember that when a man makes a sperm he doles out half his genes to it. There is therefore a 50 per cent chance that the sperm which begot your sister received the gene G. If, on the other hand, you received G from your mother, exactly parallel reasoning shows that half of her eggs must have contained G; once again, the chances are 50 per cent that your sister contains G. This means that if you had 100 brothers and sisters, approximately 50 of them would contain any particular rare gene that you contain. It also means that if you have 100 rare genes, approximately 50 of them are in the body of any one of your brothers or sisters.

You can do the same kind of calculation for any degree of kinship you like. An important relationship is that between parent and child. If you have one copy of gene H, the chance that any particular one of your children has it is 50 per cent, because half your sex cells contain H, and any particular child was made from one of those sex cells. If you have one copy of gene J, the chance that your father also had J is 50 per cent, because you received half your genes from him, and half from your mother. For convenience we use an index of *relatedness*, which expresses the chance of a gene being shared between two relatives. The relatedness between two brothers is $\frac{1}{2}$, since half the genes possessed by one brother will be found in the other. This is an average figure: by the luck of the meiotic draw, it is possible for particular pairs of brothers to share more or fewer genes than this. The relatedness between parent and child is always exactly $\frac{1}{2}$.

It is rather tedious going through the calculations from first principles every time, so here is a rough and ready rule for working out the relatedness between any two individuals *A* and *B*. You may find it useful in making your will, or in interpreting apparent resemblances in your own family. It works for all simple cases, but breaks down where incestuous mating occurs, and in certain insects, as we shall see.

First identify all the *common ancestors* of *A* and *B*. For instance, the common ancestors of a pair of first cousins are their shared grandfather and grandmother. Once you have found a common ancestor, it is of course logically true that all his ancestors are common to *A* and *B* as well. However, we ignore all but the most recent common ancestors. In this sense, first cousins have only two common ancestors. If *B* is a lineal descendant of *A*, for instance his great grandson, then *A* himself is the 'common ancestor' we are looking for.

Having located the common ancestor(s) of *A* and *B*, count the *generation distance* as follows. Starting at *A*, climb up the family tree until you hit a common ancestor, and then climb down again to *B*. The total number of steps up the tree and then down again is the generation distance. For instance, if *A* is *B*'s uncle, the generation distance is 3. The common ancestor is *A*'s father (say) and *B*'s grandfather. Starting at *A* you have to climb up one generation in order to hit the common ancestor. Then to get down to *B* you have to descend two generations on the other side. Therefore the generation distance is $1 + 2 = 3$.

Having found the generation distance between *A* and *B* via a particular common ancestor, calculate that part of their relatedness for which that ancestor is responsible. To do this, multiply $\frac{1}{2}$ by itself once for each step of the generation distance. If the generation distance is 3, this means calculate $\frac{1}{2} \times \frac{1}{2} \times \frac{1}{2}$ or $(\frac{1}{2})^3$. If the generation distance via a particular ancestor is equal to *g* steps, the portion of relatedness due to that ancestor is $(\frac{1}{2})^g$.

But this is only part of the relatedness between *A* and *B*. If they have more than one common ancestor we have to add on the equivalent figure for each ancestor. It is usually the case that the generation distance is the same for all common ancestors of a pair of individuals. Therefore, having worked out the relatedness between *A* and *B* due to any one of the ancestors, all you have to do in practice is to multiply by the number of ancestors. First cousins, for instance, have two common ancestors, and the generation distance via each one is 4. Therefore their relatedness is $2 \times (\frac{1}{2})^4 = \frac{1}{8}$. If A is B's great-grandchild, the generation distance is 3 and the number of common 'ancestors' is 1 (B himself), so the relatedness is $1 \times (\frac{1}{2})^3 = \frac{1}{8}$. Genetically speaking, your first cousin is equivalent to a great-grandchild. Similarly, you are just as likely to 'take after' your uncle (relatedness $= 2 \times (\frac{1}{2})^3 = \frac{1}{4}$) as after your grandfather (relatedness $= 1 \times (\frac{1}{2})^2 = \frac{1}{4}$).

For relationships as distant as third cousin ($2 \times (\frac{1}{2})^8 = \frac{1}{128}$), we are getting down near the baseline probability that a particular gene possessed by A will be shared by any random individual taken from the population. A third cousin is not far from being equivalent to any old Tom, Dick, or Harry as far as an altruistic gene is concerned. A second cousin (relatedness $= \frac{1}{32}$) is only a little bit special; a first cousin somewhat more so ($\frac{1}{8}$). Full brothers and sisters, and parents and children are very special ($\frac{1}{2}$), and identical twins (relatedness $= 1$) just as special as oneself. Uncles and aunts, nephews and nieces, grandparents and grandchildren, and half brothers and half sisters, are intermediate with a relatedness of $\frac{1}{4}$.

Now we are in a position to talk about genes for kin-altruism much more precisely. A gene for suicidally saving five cousins would not become more numerous in the population, but a gene for saving five brothers or ten first cousins would. The minimum requirement for a suicidal altruistic gene to be successful is that it should save more than two siblings (or children or parents), or more than four half-siblings (or uncles, aunts, nephews, nieces, grandparents, grandchildren), or more than eight first cousins, etc. Such a gene, on average, tends to live on in the bodies of enough individuals saved by the altruist to compensate for the death of the altruist itself.

If an individual could be sure that a particular person was his identical twin, he should be exactly as concerned for his twin's welfare as for his own. Any gene for twin altruism is bound to be carried by both twins, therefore if one dies heroically to save the other the gene lives on. Nine-banded armadillos are born in a litter of identical quadruplets. As far as I know, no feats of heroic self-sacrifice have been reported for young armadillos, but it has been pointed out that some strong altruism is definitely to be expected, and it would be well worth somebody's while going out to South America to have a look.

We can now see that parental care is just a special case of kin altruism. Genetically speaking, an adult should devote just as much care and attention to its orphaned baby brother as it does to one of its own children. Its relatedness to both infants is exactly the same, $\frac{1}{2}$. In gene selection terms, a gene for big sister altruistic behaviour should have just as good a chance of spreading through the population as a gene for parental altruism. In practice, this is an over-simplification for various reasons which we shall come to later, and brotherly or sisterly care is nothing like so common in nature as parental care. But the point I am making here is that there is nothing special *genetically* speaking about the parent/child relationship as against the brother/sister relationship. The fact that parents actually hand on genes to children, but sisters do not hand on genes to each other is irrelevant, since the sisters both receive identical replicas of the same genes from the same parents.

Some people use the term *kin selection* to distinguish this kind of natural selection from group selection (the differential survival of groups) and individual selection (the differential survival of individuals). Kin selection accounts for within-family altruism; the closer the relationship, the stronger the selection. There is nothing wrong with this term, but unfortunately it may have to be abandoned because of recent gross misuses of it, which are likely to muddle and confuse biologists for years to come. E. O. Wilson, in his otherwise admirable *Sociobiology: The New Synthesis*, defines kin selection as a special case of group selection. He has a diagram which clearly shows that he thinks of it as intermediate between 'individual selection', and 'group selection' in the conventional sense—the sense which I used in Chapter 1. Now group selection—even by Wilson's own definition—means the differential survival of *groups* of individuals. There is, to be sure, a sense in which a family is a special kind of group. But the whole point of Hamilton's argument is that the distinction between family and non-family is not hard and fast, but a matter of mathematical probability. It is no part of Hamilton's theory that animals should behave altruistically towards all 'members of the family', and selfishly to everybody else. There are no definite lines to be drawn between family and non-family. We do not have to decide whether, say, second cousins should count as inside the family group or outside it: we simply expect that second cousins should tend to receive $\frac{1}{16}$ as much altruism as offspring or siblings. Kin selection is emphatically *not* a special case of group selection. It is a special consequence of gene selection.

There is an even more serious shortcoming in Wilson's definition of kin selection. He deliberately excludes offspring: they don't count as kin! Now of course he knows perfectly well that offspring are kin to their parents, but he prefers not to invoke the theory of kin selection in order to explain altruistic care by parents of their own offspring. He is, of course, entitled to define a word however he likes, but this is a most confusing definition, and I hope that Wilson will change it in future editions of his justly influential book. Genetically speaking, parental care and brother/sister altruism evolve for exactly the same reason: in both cases there is a good chance that the altruistic gene is present in the body of the beneficiary.

I ask the general reader's indulgence for this little diatribe, and return hastily to the main story. So far, I have over-simplified somewhat, and it is now time to introduce some qualifications. I have talked in elemental terms of suicidal genes for saving the lives of particular numbers of kin of exactly known relatedness. Obviously, in real life, animals cannot be expected to count exactly how many relatives they are saving, nor to perform Hamilton's calculations in their heads even if they had some way of knowing exactly who their brothers and cousins were. In real life, certain suicide and absolute 'saving' of life must be replaced by *statistical risks* of death, one's own and other people's. Even a third cousin may be worth saving, if the risk to yourself is very small. Then again, both you and the relative you are thinking of saving are going to die one day in any case. Every individual has an 'expectation of life' which an actuary could calculate with a certain probability of error. To save the life of a relative who is soon going to die of old age has less of an impact on the gene pool of the future than to save the life of an equally close relative who has the bulk of his life ahead of him.

Our neat symmetrical calculations of relatednesses have to be modified by messy actuarial weightings. Grandparents and grandchildren have, genetically speaking, equal reason to behave altruistically to each other, since they share $\frac{1}{4}$ of each other's genes. But if the grandchildren have the greater expectation of life, genes for grandparent to grandchild altruism have a higher selective advantage than genes for grandchild to grandparent altruism. It is quite possible for the net benefit of assisting a young distant relative to exceed the net benefit of assisting an old close relative. (Incidentally, it is not, of course, necessarily the case that grandparents have a shorter expectation of life than grandchildren. In species with a high infant-mortality rate, the reverse may be true.)

To extend the actuarial analogy, individuals can be thought of as life-insurance underwriters. An individual can be expected to invest or risk a certain proportion of his own assets in the life of another individual. He takes into account his relatedness to the other individual, and also whether the individual is a 'good risk' in terms of his life expectancy compared with the insurer's own. Strictly we should say 'reproduction expectancy' rather than 'life expectancy', or to be even more strict, 'general capacity to benefit own genes in the future expectancy'. Then in order for altruistic behaviour to evolve, the net risk to the altruist must be less than the net benefit to the recipient multiplied by the relatedness. Risks and benefits have to be calculated in the complex actuarial way I have outlined.

But what a complicated calculation to expect a poor survival machine to do, especially in a hurry! Even the great mathematical biologist J. B. S. Haldane (in a paper of 1955 in which he anticipated Hamilton by postulating the spread of a gene for saving close relatives from drowning) remarked: '. . . on the two occasions when I have pulled possibly drowning people out of the water (at an infinitesimal risk to myself) I had no time to make such calculations.' Fortunately, however, as Haldane well knew, it is not necessary to assume that survival machines do the sums consciously in their heads. Just as we may use a slide rule without appreciating that we are, in effect, using logarithms, so an animal may be pre-programmed in such a way that it behaves *as if* it had made a complicated calculation.

This is not so difficult to imagine as it appears. When a man throws a ball high in the air and catches it again, he behaves as if he had solved a set of differential equations in predicting the trajectory of the ball. He may neither know nor care what a differential equation is, but this does not affect his skill with the ball. At some subconscious level, something functionally equivalent to the mathematical calculations is going on. Similarly, when a man takes a difficult decision, after weighing up all the pros and cons, and all the consequences of the decision which he can imagine, he is doing the functional equivalent of a large 'weighted sum' calculation, such as a computer might perform.

If we were to program a computer to simulate a model survival machine making decisions about whether to behave altruistically, we should probably proceed roughly as follows. We should make a list of all the alternative things the animal might do. Then for each of these alternative behaviour patterns we program a weighted sum calculation. All the various benefits will have a plus sign; all the risks will have a minus sign; both benefits and risks will be *weighted* by being multiplied by the appropriate index of relatedness before being added up. For simplicity we can, to begin with, ignore other weightings, such as those for age and health. Since an individual's 'relatedness' with himself is 1 (i.e. he has 100 per cent of his own genes—obviously), risks and benefits to himself will not be devalued at all, but will be given their full weight in the calculation. The whole sum for any one of the alternative behaviour patterns will look like this: Net benefit of behaviour pattern =   Benefit to self − Risk to self + $\frac{1}{2}$ Benefit to brother − $\frac{1}{2}$ Risk to brother + $\frac{1}{2}$ Benefit to other brother − $\frac{1}{2}$ Risk to other brother + $\frac{1}{8}$ Benefit to first cousin − $\frac{1}{8}$ Risk to first cousin + $\frac{1}{2}$ Benefit to child − $\frac{1}{2}$ Risk to child + etc.

The result of the sum will be a number called the net benefit score of that behaviour pattern. Next, the model animal computes the equivalent sum for each alternative behaviour pattern in his repertoire. Finally he chooses to perform the behaviour pattern which emerges with the largest net benefit. Even if all the scores come out negative, he should still choose the action with the highest one, the least of evils. Remember that any positive action involves consumption of energy and time, both of which could have been spent doing other things. If doing nothing emerges as the 'behaviour' with the highest net benefit score, the model animal will do nothing.

Here is a very over-simplified example, this time expressed in the form of a subjective soliloquy rather than a computer simulation. I am an animal who has found a clump of eight mushrooms. After taking account of their nutritional value, and subtracting something for the slight risk that they might be poisonous, I estimate that they are worth +6 units each (the units are arbitrary pay-offs as in the previous chapter). The mushrooms are so big I could eat only three of them. Should I inform anybody else about my find, by giving a 'food call'? Who is within earshot? Brother *B* (his relatedness to me is $\frac{1}{2}$), cousin *C* (relatedness to me = $\frac{1}{8}$), and *D* (no particular relation: his relatedness to me is some small number which can be treated as zero for practical purposes). The net benefit score to me if I keep quiet about my find will be +6 for each of the three mushrooms I eat, that is +18 in all. My net benefit score if I give the food call needs a bit of figuring. The eight mushrooms will be shared equally between the four of us. The pay-off to me from the two that I eat myself will be the full +6 units each, that is +12 in all. But I shall also get some pay-off when my brother and cousin eat their two mushrooms each, because of our shared genes. The actual score comes to $(1 \times 12) + (\frac{1}{2} \times 12) + (\frac{1}{8} \times 12) + (0 \times 12) = +19\frac{1}{2}$. The corresponding net benefit for the selfish behaviour was +18: it is a close-run thing, but the verdict is clear. I should give the food-call; altruism on my part would in this case pay my selfish genes.

I have made the simplifying assumption that the individual animal works out what is best for his genes. What really happens is that the gene pool becomes filled with genes which influence bodies in such a way that they behave as if they had made such calculations.

In any case the calculation is only a very preliminary first approximation to what it ideally should be. It neglects many things, including the ages of the individuals concerned. Also, if I have just had a good meal, so that I can only find room for one mushroom, the net benefit of giving the food call will be greater than it would be if I was famished. There is no end to the progressive refinements of the calculation which could be achieved in the best of all possible worlds. But real life is not lived in the best of all possible worlds. We cannot expect real animals to take every last detail into account in coming to an optimum decision. We shall have to discover, by observation and experiment in the wild, how closely real animals actually come to achieving an ideal cost–benefit analysis.

Just to reassure ourselves that we have not become too carried away with subjective examples, let us briefly return to gene language. Living bodies are machines programmed by genes who have survived. The genes who have survived have done so in conditions which tended *on average* to characterize the environment of the species in the past. Therefore 'estimates' of costs and benefits are based on past 'experience', just as they are in human decision-making. However, experience in this case has the special meaning of gene experience or, more precisely, conditions of past gene survival. (Since genes also endow survival machines with the capacity to learn, some cost–benefit estimates could be said to be taken on the basis of individual experience as well.) So long as conditions do not change too drastically, the estimates will be good estimates, and survival machines will tend to make the right decisions on average. If conditions change radically, survival machines will tend to make erroneous decisions, and their genes will pay the penalty. Just so; human decisions based on outdated information tend to be wrong.

Estimates of relatedness are also subject to error and uncertainty. In our over-simplified calculations so far, we have talked as if survival machines *know* who is related to them, and how closely. In real life such certain knowledge is occasionally possible, but more usually the relatedness can only be estimated as an average number. For example, suppose that A and B could equally well be either half brothers or full brothers. Their relatedness is either $\frac{1}{4}$ or $\frac{1}{2}$, but since we do not know whether they are half or full brothers, the effectively usable figure is the average, $\frac{3}{8}$. If it is certain that they have the same mother, but the odds that they have the same father are only 1 in 10, then it is 90 per cent certain that they are half brothers, and 10 per cent certain that they are full brothers, and the effective relatedness is $\frac{1}{10} \times \frac{1}{2} + \frac{9}{10} \times \frac{1}{4} = 0.275$.

But when we say something like 'it' is 90 per cent certain, what 'it' are we referring to? Do we mean a human naturalist after a long field study is 90 per cent certain, or do we mean the animals are 90 per cent certain? With a bit of luck these two may amount to nearly the same thing. To see this, we have to think how animals might actually go about estimating who their close relations are.

We know who our relations are because we are told, because we give them names, because we have formal marriages, and because we have written records and good memories. Many social anthropologists are preoccupied with 'kinship' in the societies which they study. They do not mean real genetic kinship, but subjective and cultural ideas of kinship. Human customs and tribal rituals commonly give great emphasis to kinship; ancestor worship is widespread, family obligations and loyalties dominate much of life. Blood-feuds and inter-clan warfare are easily interpretable in terms of Hamilton's genetic theory. Incest taboos testify to the great kinship-consciousness of man, although the genetical advantage of an incest taboo is nothing to do with altruism; it is presumably concerned with the injurious effects of recessive genes which appear with inbreeding. (For some reason many anthropologists do not like this explanation.)

How could wild animals 'know' who their kin are, or in other words, what behavioural rules could they follow which would have the indirect effect of making them seem to know about kinship? The rule 'be nice to your relations' begs the question of how relations are to be recognized in practice. Animals have to be given by their genes a simple rule for action, a rule that does not involve all-wise cognition of the ultimate purpose of the action, but a rule which works nevertheless, at least in average conditions. We humans are familiar with rules, and so powerful are they that if we are small minded we obey a rule itself, even when we can see perfectly well that it is not doing us, or anybody else, any good. For instance, some orthodox Jews and Muslims would starve rather than break their rule against eating pork. What simple practical rules could animals obey which, under normal conditions, would have the indirect effect of benefiting their close relations?

If animals had a tendency to behave altruistically towards individuals who physically resembled them, they might indirectly be doing their kin a bit of good. Much would depend on details of the species concerned. Such a rule would, in any case, only lead to 'right' decisions in a statistical sense. If conditions changed, for example if a species started living in much larger groups, it could lead to wrong decisions. Conceivably, racial prejudice could be interpreted as an irrational generalization of a kin-selected tendency to identify with individuals physically resembling oneself, and to be nasty to individuals different in appearance.

In a species whose members do not move around much, or whose members move around in small groups, the chances may be good that any random individual you come across is fairly close kin to you. In this case the rule 'Be nice to any member of the species whom you meet' could have positive survival value, in the sense that a gene predisposing its possessors to obey the rule might become more numerous in the gene pool. This may be why altruistic behaviour is so frequently reported in troops of monkeys and schools of whales. Whales and dolphins drown if they are not allowed to breathe air. Baby whales, and injured individuals who cannot swim to the surface, have been seen to be rescued and held up by companions in the school. It is not known whether whales have ways of knowing who their close relatives are, but it is possible that it does not matter. It may be that the overall probability that a random member of the school is a relation is so high that the altruism is worth the cost. Incidentally, there is at least one well-authenticated story of a drowning human swimmer being rescued by a wild dolphin. This could be regarded as a misfiring of the rule for saving drowning members of the school. The rule's 'definition' of a member of the school who is drowning might be something like: 'A long thing thrashing about and choking near the surface.'

Adult male baboons have been reported to risk their lives defending the rest of the troop against predators such as leopards. It is quite probable that any adult male has, on average, a fairly large number of genes tied up in other members of the troop. A gene which 'says', in effect: 'Body, if you happen to be an adult male, defend the troop against leopards', could become more numerous in the gene pool. Before leaving this often-quoted example, it is only fair to add that at least one respected authority has reported very different facts. According to her, adult males are the first over the horizon when a leopard appears.

Baby chicks feed in family clutches, all following their mother. They have two main calls. In addition to the loud piercing cheep which I have already mentioned, they give short melodious twitters when feeding. The cheeps, which have the effect of summoning the mother's aid, are ignored by the other chicks. The twitters, however, are attractive to chicks. This means that when one chick finds food, its twitters attract other chicks to the food as well: in the terms of the earlier hypothetical example, the twitters are 'food calls'. As in that case, the apparent altruism of the chicks can easily be explained by kin selection. Since, in nature, the chicks would all be full brothers and sisters, a gene for giving the food twitter would spread, provided the cost to the twitterer is less than twice the net benefit to the other chicks. As the benefit is shared out between the whole clutch, which normally numbers more than two, it is not difficult to imagine this condition being realized. Of course the rule misfires in domestic or farm situations when a hen is made to sit on eggs not her own, even turkey or duck eggs. But neither the hen nor her chicks can be expected to realize this. Their behaviour has been shaped under the conditions which normally prevail in nature, and in nature strangers are not normally found in your nest.

Mistakes of this sort may, however, occasionally happen in nature. In species which live in herds or troops, an orphaned youngster may be adopted by a strange female, most probably one who has lost her own child. Monkey-watchers sometimes use the word 'aunt' for an adopting female. In most cases there is no evidence that she really is an aunt, or indeed any kind of relative: if monkey-watchers were as gene-conscious as they might be, they would not use an important word like 'aunt' so uncritically. In most cases we should probably regard adoption, however touching it may seem, as a misfiring of a built-in rule. This is because the generous female is doing her own genes no good by caring for the orphan. She is wasting time and energy which she could be investing in the lives of her own kin, particularly future children of her own. It is presumably a mistake which happens too seldom for natural selection to have 'bothered' to change the rule by making the maternal instinct more selective. In many cases, by the way, such adoptions do not occur, and an orphan is left to die.

There is one example of a mistake which is so extreme that you may prefer to regard it not as a mistake at all, but as evidence against the selfish gene theory. This is the case of bereaved monkey mothers who have been seen to steal a baby from another female, and look after it. I see this as a double mistake, since the adopter not only wastes her own time; she also releases a rival female from the burden of child-rearing, and frees her to have another child more quickly. It seems to me a critical example which deserves some thorough research. We need to know how often it happens; what the average relatedness between adopter and child is likely to be; and what the attitude of the real mother of the child is—it is, after all, to her advantage that her child *should* be adopted; do mothers deliberately try to deceive naïve young females into adopting their children? (It has also been suggested that adopters and baby-snatchers might benefit by gaining valuable practice in the art of child-rearing.)

An example of a deliberately engineered misfiring of the maternal instinct is provided by cuckoos, and other 'brood-parasites'—birds who lay their eggs in somebody else's nest. Cuckoos exploit the rule built into bird parents: 'Be nice to any small bird sitting in the nest which you built.' Cuckoos apart, this rule will normally have the desired effect of restricting altruism to immediate kin, because it happens to be a fact that nests are so isolated from each other that the contents of your own nest are almost bound to be your own chicks. Adult herring gulls do not recognize their own eggs, and will happily sit on other gull eggs, and even crude wooden dummies if these are substituted by a human experimenter. In nature, egg recognition is not important for gulls, because eggs do not roll far enough to reach the vicinity of a neighbour's nest, some yards away. Gulls do, however, recognize their own chicks: chicks, unlike eggs, wander, and can easily end up near the nest of a neighbouring adult, often with fatal results, as we saw in Chapter 1.

Guillemots, on the other hand, do recognize their own eggs by means of the speckling pattern, and actively discriminate in favour of them when incubating. This is presumably because they nest on flat rocks, where there is a danger of eggs rolling around and getting muddled up. Now, it might be said, why do they bother to discriminate and sit only on their own eggs? Surely if everybody saw to it that she sat on somebody's egg, it would not matter whether each particular mother was sitting on her own or somebody else's. This is the argument of a group selectionist. Just consider what would happen if such a group baby-sitting circle did develop. The average clutch size of the guillemot is one. This means that if the mutual baby-sitting circle is to work successfully, every adult would have to sit on an average of one egg. Now suppose somebody cheated, and refused to sit on an egg. Instead of wasting time sitting, she could spend her time laying more eggs. And the beauty of the scheme is that the other, more altruistic, adults would look after them for her. They would go on faithfully obeying the rule 'If you see a stray egg near your nest, haul it in and sit on it.' So the gene for cheating the system would spread through the population, and the nice friendly baby-sitting circle would break down.

'Well', it might be said, 'what if the honest birds retaliated by refusing to be blackmailed, and resolutely decided to sit on one egg and only one egg? That should foil the cheaters, because they would see their own eggs lying out on the rocks with nobody incubating them. That should soon bring them into line.' Alas, it would not. Since we are postulating that the sitters are not discriminating one egg from another, if the honest birds put into practice this scheme for resisting cheating, the eggs which ended up being neglected would be just as likely to be their own eggs as those of the cheaters. The cheaters would still have the advantage, because they would lay more eggs and have more surviving children. The only way an honest guillemot could beat the cheaters would be to discriminate actively in favour of her own eggs. That is, to cease being altruistic and look after her own interests.

To use the language of Maynard Smith, the altruistic adoption 'strategy' is not an evolutionarily stable strategy. It is unstable in the sense that it can be bettered by a rival selfish strategy of laying more than one's fair share of eggs, and then refusing to sit on them. This latter selfish strategy is in its turn unstable, because the altruistic strategy which it exploits is unstable, and will disappear. The only evolutionarily stable strategy for a guillemot is to recognize its own egg, and sit exclusively on its own egg, and this is exactly what happens.

The song-bird species which are parasitized by cuckoos have

fought back, not in this case by learning the appearance of their own eggs, but by discriminating instinctively in favour of eggs with the species-typical markings. Since they are not in danger of being parasitized by members of their own species, this is effective. But the cuckoos have retaliated in their turn by making their eggs more and more like those of the host species in colour, size, and markings. This is an example of a lie, and it often works. The result of this evolutionary arms race has been a remarkable perfection of mimicry on the part of the cuckoo eggs. We may suppose that a proportion of cuckoo eggs and chicks are 'found out', and those which are not found out are the ones who live to lay the next generation of cuckoo eggs. So genes for more effective deception spread through the cuckoo gene pool. Similarly, those host birds with eyes sharp enough to detect any slight imperfection in the cuckoo eggs' mimicry are the ones who contribute most to their own gene pool. Thus sharp and sceptical eyes are passed on to their next generation. This is a good example of how natural selection can sharpen up active discrimination, in this case discrimination against another species whose members are doing their best to foil the discriminators.

Now let us return to the comparison between an animal's 'estimate' of its kinship with other members of its group, and the corresponding estimate of an expert field naturalist. Brian Bertram has spent many years studying the biology of lions in the Serengeti National Park. On the basis of his knowledge of their reproductive habits, he has estimated the average relatedness between individuals in a typical lion pride. The facts which he uses to make his estimates are things like this. A typical pride consists of seven adult females who are its more permanent members, and two adult males who are itinerant. About half the adult females give birth as a batch at the same time, and rear their cubs together so that it is difficult to tell which cub belongs to whom. The typical litter size is three cubs. The fathering of litters is shared equally between the adult males in the pride. Young females remain in the pride, and replace old females who die or leave. Young males are driven out when adolescent. When they grow up, they wander around from pride to pride in small related gangs or pairs, and are unlikely to return to their original family.

Using these and other assumptions, you can see that it would be possible to compute an average figure for the relatedness of two individuals from a typical lion pride. Bertram arrives at a figure of 0·22 for a pair of randomly chosen males, and 0·15 for a pair of females. That is to say, males within a pride are on average slightly less close than half brothers, and females slightly closer than first cousins.

Now, of course, any particular pair of individuals might be full brothers, but Bertram had no way of knowing this, and it is a fair bet that the lions did not know it either. On the other hand, the average figures which Bertram estimated are available to the lions themselves in a certain sense. If these figures really are typical for an average lion pride, then any gene which predisposed males to behave towards other males as if they were nearly half brothers would have positive survival value. Any gene which went too far, and made males behave in a friendly way more appropriate to full brothers would on average be penalized, as would a gene for not being friendly enough, say treating other males like second cousins. If the facts of lion life are as Bertram says, and, just as important, if they have been like that for a large number of generations, then we may expect that natural selection will have favoured a degree of altruism appropriate to the average degree of relatedness in a typical pride. This is what I meant when I said that the kinship estimates of animal and of good naturalist might end up rather the same.

So we conclude that the 'true' relatedness may be less important in the evolution of altruism than the best *estimate* of relatedness that animals can get. This fact is probably a key to understanding why parental care is so much more common and more devoted than brother/sister altruism in nature, and also why animals may value themselves more highly even than several brothers. Briefly, what I am saying is that, in addition to the index of relatedness, we should consider something like an index of 'certainty'. Although the parent/child relationship is no closer genetically than the brother/sister relationship, its certainty is greater. It is normally possible to be much more certain who your children are than who your brothers are. And you can be more certain still who you yourself are!

We considered cheaters among guillemots, and we shall have

more to say about liars and cheaters and exploiters in following chapters. In a world where other individuals are constantly on the alert for opportunities to exploit kin-selected altruism, and use it for their own ends, a survival machine has to consider who it can trust, who it can be really sure of. *If B* is really my baby brother, then I should care for him up to half as much as I care for myself, and fully as much as I care for my own child. But can I be as sure of him as I can of my own child? How do I know he is my baby brother?

If C is my identical twin, then I should care for him twice as much as I care for any of my children, indeed I should value his life no less than my own. But can I be sure of him? He looks like me to be sure, but it could be that we just happen to share the genes for facial features. No, I will not give up my life for him, because although it is *possible* that he bears 100 per cent of my genes, I absolutely *know* that I contain 100 per cent of my genes, so I am worth more to me than he is. I am the only individual that any one of my selfish genes can be sure of. And although ideally a gene for individual selfishness could be displaced by a rival gene for altruistically saving at least one identical twin, two children or brothers, or at least four grandchildren etc., the gene for individual selfishness has the enormous advantage of *certainty* of individual identity. The rival kin-altruistic gene runs the risk of making mistakes of identity, either genuinely accidental, or deliberately engineered by cheats and parasites. We therefore must expect individual selfishness in nature, to an extent greater than would be predicted by considerations of genetic relatedness alone.

In many species a mother can be more sure of her young than a father can. The mother lays the visible, tangible egg, or bears the child. She has a good chance of knowing for certain the bearers of her own genes. The poor father is much more vulnerable to deception. It is therefore to be expected that fathers will put less effort than mothers into caring for young. We shall see that there are other reasons to expect the same thing, in the chapter on the Battle of the Sexes (Chapter 9). Similarly, maternal grandmothers can be more sure of their grandchildren than paternal grandmothers can, and might be expected to show more altruism than paternal grandmothers. This is because they can be sure of their daughter's children, but their son may have been cuckolded. Maternal grandfathers are just as sure of their grandchildren as paternal grandmothers are, since both can reckon on one generation of certainty and one generation of uncertainty. Similarly, uncles on the mother's side should be more interested in the welfare of nephews and nieces than uncles on the father's side, and in general should be just as altruistic as aunts are. Indeed in a society with a high degree of marital infidelity, maternal uncles should be more altruistic than 'fathers' since they have more grounds for confidence in their relatedness to the child. They know that the child's mother is at least their half-sister. The 'legal' father knows nothing. I do not know of any evidence bearing on these predictions, but I offer them in the hope that others may, or may start looking for evidence. In particular, perhaps social anthropologists might have interesting things to say.

Returning to the fact that parental altruism is more common than fraternal altruism, it does seem reasonable to explain this in terms of the 'identification problem'. But this does not explain the fundamental asymmetry in the parent/child relationship itself. Parents care more for their children than children do for their parents, although the genetic relationship is symmetrical, and certainty of relatedness is just as great both ways. One reason is that parents are in a better practical position to help their young, being older and more competent at the business of living. Even if a baby wanted to feed its parents, it is not well equipped to do so in practice.

There is another asymmetry in the parent/child relationship which does not apply to the brother/sister one. Children are always younger than their parents. This often, though not always, means they have a longer expectation of life. As I emphasized above, expectation of life is an important variable which, in the best of all possible worlds, should enter into an animal's 'calculation' when it is 'deciding' whether to behave altruistically or not. In a species in which children have a longer average life-expectancy than parents, any gene for child altruism would be labouring under a disadvantage. It would be engineering altruistic self-sacrifice for the benefit of individuals who are nearer to dying of old age than the altruist itself. A gene for parent altruism, on the other hand, would have a corresponding advantage as far as the life-expectancy terms in the equation were concerned.

One sometimes hears it said that kin selection is all very well as a theory, but there are few examples of its working in practice. This criticism can only be made by someone who does not understand what kin selection means. The truth is that all examples of child-protection and parental care, and all associated bodily organs, milk-secreting glands, kangaroo pouches, and so on, are examples of the working in nature of the kin-selection principle. The critics are of course familiar with the widespread existence of parental care, but they fail to understand that parental care is no less an example of kin selection than brother/sister altruism. When they say they want examples, they mean that they want examples other than parental care, and it is true that such examples are less common. I have suggested reasons why this might be so. I could have gone out of my way to quote examples of brother/sister altruism—there are in fact quite a few. But I don't want to do this, because it would reinforce the erroneous idea (favoured, as we have seen, by Wilson) that kin selection is specifically about relationships *other than* the parent/child relationship.

The reason this error has grown up is largely historical. The evolutionary advantage of parental care is so obvious that we did not have to wait for Hamilton to point it out. It has been understood ever since Darwin. When Hamilton demonstrated the genetic equivalence of other relationships, and their evolutionary significance, he naturally had to lay stress on these other relationships. In particular, he drew examples from the social insects such as ants and bees, in which the sister/sister relationship is particularly important, as we shall see in a later chapter. I have even heard people say that they thought Hamilton's theory applied *only* to the social insects!

If anybody does not want to admit that parental care is an example of kin selection in action, then the onus is on him to formulate a general theory of natural selection which predicts parental altruism, but which does *not* predict altruism between collateral kin. I think he will fail.

7. Family planning

It is easy to see why some people have wanted to separate parental care from the other kinds of kin-selected altruism. Parental care looks like an integral part of reproduction whereas, for example, altruism toward a nephew is not. I think there really is an important distinction hidden here, but that people have mistaken what the distinction is. They have put reproduction and parental care on one side, and other sorts of altruism on the other. But I wish to make a distinction between *bringing new individuals into the world*, on the one hand, and *caring for existing individuals* on the other. I shall call these two activities respectively child-bearing and child-caring. An individual survival machine has to make two quite different sorts of decisions, caring decisions and bearing decisions. I use the word decision to mean unconscious strategic move. The caring decisions are of this form: 'There is a child; its degree of relatedness to me is so and so; its chances of dying if I do not feed it are such and such; shall I feed it?' Bearing decisions, on the other hand, are like this: 'Shall I take whatever steps are necessary in order to bring a new individual into the world; shall I reproduce?' To some extent, caring and bearing are bound to compete with each other for an individual's time and other resources: the individual may have to make a choice: 'Shall I care for this child or shall I bear a new one?'

Depending on the ecological details of the species, various mixes of caring and bearing strategies can be evolutionarily stable. The one thing which cannot be evolutionarily stable is a *pure* caring strategy. If all individuals devoted themselves to caring for existing children to such an extent that they never brought any new ones into the world, the population would quickly become invaded by mutant individuals who specialized in bearing. Caring can only be evolutionarily stable as part of a mixed strategy—at least some bearing has to go on.

The species with which we are most familiar—mammals and

birds—tend to be great carers. A decision to bear a new child is usually followed by a decision to care for it. It is because bearing and caring so often go together in practice that people have muddled the two things up. But from the point of view of the selfish genes there is, as we have seen, no distinction in principle between caring for a baby brother and caring for a baby son. Both infants are equally closely related to you. If you have to choose between feeding one or the other, there is no genetic reason why you should choose your own son. But on the other hand you cannot, by definition, bear a baby brother. You can only care for him once somebody else has brought him into the world. In the last chapter we looked at how individual survival machines ideally should decide whether to behave altruistically towards other individuals who already exist. In this chapter we look at how they should decide whether to bring new individuals into the world.

It is over this matter that the controversy about 'group selection', which I mentioned in Chapter 1, has chiefly raged. This is because Wynne-Edwards, who has been mainly responsible for promulgating the idea of group selection, did so in the context of a theory of 'population regulation'. He suggested that individual animals deliberately and altruistically reduce their birth rates for the good of the group as a whole.

This is a very attractive hypothesis, because it fits so well with what individual humans ought to do. Mankind is having too many children. Population size depends upon four things: births, deaths, immigrations, and emigrations. Taking the world population as a whole, immigrations and emigrations do not occur, and we are left with births and deaths. So long as the average number of children per couple is larger than two surviving to reproduce, the numbers of babies born will tend to increase over the years at an ever-accelerating rate. In each generation the population, instead of going up by a fixed amount, increases by something more like a fixed proportion of the size that it has already reached. Since this size is itself getting bigger, the size of the increment gets bigger. If this kind of growth was allowed to go on unchecked, a population would reach astronomical proportions surprisingly quickly.

Incidentally, a thing which is sometimes not realized even by people who worry about population problems is that population growth depends on *when* people have children, as well as on how many they have. Since populations tend to increase by a certain proportion *per generation*, if follows that if you space the generations out more, the population will grow at a slower rate per year. Banners which read 'Stop at Two' could equally well be changed to 'Start at Thirty'! But in any case, accelerating population growth spells serious trouble.

We have probably all seen examples of the startling calculations which can be used to bring this home. For instance, the present population of Latin America is around 300 million, and already many of them are under-nourished. But if the population continued to increase at the present rate, it would take less than 500 years to reach the point where the people, packed in a standing position, formed a solid human carpet over the whole area of the continent. This is so, even if we assume them to be very skinny—a not unrealistic assumption. In 1000 years from now they would be standing on each other's shoulders more than a million deep. By 2000 years, the mountain of people, travelling outwards at the speed of light, would have reached the edge of the known universe.

It will not have escaped you that this is a hypothetical calculation! It will not really happen like that for some very good practical reasons. The names of some of these reasons are famine, plague, and war; *or*, if we are lucky, birth control. It is no use appealing to advances in agricultural science—'green revolutions' and the like. Increases in food production may temporarily alleviate the problem, but it is mathematically certain that they cannot be a long-term solution; indeed, like the medical advances which have precipitated the crisis, they may well make the problem worse, by speeding up the rate of the population expansion. It is a simple logical truth that, short of mass emigration into space, with rockets taking off at the rate of several million per second, uncontrolled birth-rates are bound to lead to horribly increased death-rates. It is hard to believe that this simple truth is not understood by those leaders who forbid their followers to use effective contraceptive methods. They express a preference for 'natural' methods of population limitation, and a natural method is exactly what they are going to get. It is called starvation.

But of course the unease which such long-term calculations

arouse is based on concern for the future welfare of our species as a whole. Humans (some of them) have the conscious foresight to see ahead to the disastrous consequences of over-population. It is the basic assumption of this book that survival machines in general are guided by selfish genes, who most certainly cannot be expected to see into the future, nor to have the welfare of the whole species at heart. This is where Wynne-Edwards parts company with orthodox evolutionary theorists. He thinks there is a way in which genuine altruistic birth-control can evolve.

A point that is not emphasized in the writings of Wynne-Edwards, or in Ardrey's popularization of his views, is that there is a large body of agreed facts which are not in dispute. It is an obvious fact that wild animal populations do not grow at the astronomical rates of which they are theoretically capable. Sometimes wild animal populations remain rather stable, with birth-rates and death-rates roughly keeping pace with each other. In many cases, lemmings being a famous example, the population fluctuates widely, with violent explosions alternating with crashes and near extinction. Occasionally the result is outright extinction, at least of the population in a local area. Sometimes, as in the case of the Canadian lynx—where estimates are obtained from the numbers of pelts sold by the Hudson's Bay Company in successive years—the population seems to oscillate rhythmically. The one thing animal populations do not do is go on increasing indefinitely.

Wild animals almost never die of old age: starvation, disease, or predators catch up with them long before they become really senile. Until recently this was true of man too. Most animals die in childhood, many never get beyond the egg stage. Starvation and other causes of death are the ultimate reasons why populations cannot increase indefinitely. But as we have seen for our own species, there is no necessary reason why it ever has to come to that. If only animals would regulate their *birth-rates*, starvation need never happen. It is Wynne-Edwards's thesis that that is exactly what they do. But even here there is less disagreement than you might think from reading his book. Adherents of the selfish gene theory would readily agree that animals *do* regulate their birth-rates. Any given species tends to have a rather fixed clutch-size or litter-size: no animal has an infinite number of children. The disagreement comes not over *whether* birth-rates are regulated. The disagreement is over *why* they are regulated: by what process of natural selection has family-planning evolved? In a nutshell, the disagreement is over whether animal birth-control is altruistic, practised for the good of the group as a whole; or selfish, practised for the good of the individual doing the reproducing. I will deal with the two theories in order.

Wynne-Edwards supposed that individuals have fewer children than they are capable of, for the benefit of the group as a whole. He recognized that normal natural selection cannot possibly give rise to the evolution of such altruism: the natural selection of lower-than-average reproductive rates is, on the face of it, a contradiction in terms. He therefore invoked group selection, as we saw in Chapter 1. According to him, groups whose individual members restrain their own birth-rates are less likely to go extinct than rival groups whose individual members reproduce so fast that they endanger the food supply. Therefore the world becomes populated by groups of restrained breeders. The individual restraint which Wynne-Edwards is suggesting amounts in a general sense to birth-control, but he is more specific than this, and indeed comes up with a grand conception in which the whole of social life is seen as a mechanism of population regulation. For instance, two major features of social life in many species of animals are *territoriality* and *dominance hierarchies*, already mentioned in Chapter 5.

Many animals devote a great deal of time and energy to apparently 'defending' an area of ground which naturalists call a territory. The phenomenon is very widespread in the animal kingdom, not only in birds, mammals, and fish, but in insects and even sea-anemones. The territory may be a large area of woodland which is the principle foraging ground of a breeding pair, as in the case of robins. Or, in herring gulls for instance, it may be a small area containing no food, but with a nest at its centre. Wynne-Edwards believes that animals who fight over territory are fighting over a *token* prize, rather than an actual prize like a bit of food. In many cases females refuse to mate with males who do not possess a territory. Indeed it often happens that a female whose mate is defeated and his territory conquered promptly attaches herself to the victor. Even in apparently faithful monogamous species, the female may be wedded to a male's territory rather than to him personally.

If the population gets too big, some individuals will not get territories, and therefore will not breed. Winning a territory is therefore, to Wynne-Edwards, like winning a ticket or licence to breed. Since there is a finite number of territories available, it is as if a finite number of breeding licences is issued. Individuals may fight over who gets these licences, but the total number of babies that the population can have as a whole is limited by the number of territories available. In some cases, for instance in red grouse, individuals do, at first sight, seem to show restraint, because those who cannot win territories not only do not breed; they also appear to give up the struggle to win a territory. It is as though they all accepted the rules of the game: that if, by the end of the competition season, you have not secured one of the official tickets to breed, you voluntarily refrain from breeding and leave the lucky ones unmolested during the breeding season, so that they can get on with propagating the species.

Wynne-Edwards interprets dominance hierarchies in a similar way. In many groups of animals, especially in captivity, but also in some cases in the wild, individuals learn each other's identity, and they learn whom they can beat in a fight, and who usually beats them. As we saw in Chapter 5, they tend to submit without a struggle to individuals who they 'know' are likely to beat them anyway. As a result a naturalist is able to describe a dominance hierarchy or 'peck order' (so called because it was first described for hens)—a rank-ordering of society in which everybody knows his place, and does not get ideas above his station. Of course sometimes real earnest fights do take place, and sometimes individuals can win promotion over their former immediate bosses. But as we saw in Chapter 5, the overall effect of the automatic submission by lower-ranking individuals is that few prolonged fights actually take place, and serious injuries seldom occur.

Many people think of this as a 'good thing' in some vaguely group-selectionist way. Wynne-Edwards has an altogether more daring interpretation. High-ranking individuals are more likely to breed than low-ranking individuals, either because they are preferred by females, or because they physically prevent low-ranking males from getting near females. Wynne-Edwards sees high social rank as another ticket of entitlement to reproduce. Instead of fighting directly over females themselves, individuals fight over social status, and then accept that if they do not end up high on the social scale they are not entitled to breed. They restrain themselves where females are directly concerned, though they may try every now and then to win higher status, and therefore could be said to compete *indirectly* over females. But, as in the case of territorial behaviour, the result of this 'voluntary acceptance' of the rule that only high-status males should breed is, according to Wynne-Edwards, that populations do not grow too fast. Instead of actually having too many children, and then finding out the hard way that it was a mistake, populations use formal contests over status and territory as a means of limiting their size slightly below the level at which starvation itself actually takes its toll.

Perhaps the most startling of Wynne-Edwards's ideas is that of *epideictic* behaviour, a word which he coined himself. Many animals spend a great deal of time in large flocks, herds, or shoals. Various more or less common-sense reasons why such aggregating behaviour should have been favoured by natural selection have been suggested, and I will talk about some of them in Chapter 10. Wynne-Edwards's idea is quite different. He proposes that when huge flocks of starlings mass at evening, or crowds of midges dance over a gate-post, they are performing a census of their population. Since he is supposing that individuals restrain their birth-rates in the interests of the group as a whole, and have fewer babies when the population density is high, it is reasonable that they should have some way of measuring the population density. Just so; a thermostat needs a thermometer as an integral part of its mechanism. For Wynne-Edwards, epideictic behaviour is deliberate massing in crowds to facilitate population estimation. He is not suggesting conscious population estimation, but an automatic nervous or hormonal mechanism linking the individuals' sensory perception of the density of their population with their reproductive systems.

I have tried to do justice to Wynne-Edwards's theory, even if rather briefly. If I have succeeded, you should now be feeling persuaded that it is, on the face of it, rather plausible. But the earlier chapters of this book should have prepared you to be sceptical to the point of saying that, plausible as it may sound, the evidence for Wynne-Edwards's theory had better be good, or else. . . . And unfortunately the evidence is not good. It consists of a large number of examples which could be interpreted in his way, but which could equally well be interpreted on more orthodox 'selfish gene' lines.

Although he would never have used that name, the chief architect of the selfish gene theory of family planning was the great ecologist David Lack. He worked especially on clutch-size in wild birds, but his theories and conclusions have the merit of being generally applicable. Each bird species tends to have a typical clutch size. For instance, gannets and guillemots incubate one egg at a time, swifts three, great tits half a dozen or more. There is variation in this: some swifts lay only two at a time, great tits may lay twelve. It is reasonable to suppose that the number of eggs a female lays and incubates is at least partly under genetic control, like any other characteristic. That is say there may be a gene for laying two eggs, a rival allele for laying three, another allele for laying four, and so on, although in practice it is unlikely to be quite as simple as this. Now the selfish gene theory requires us to ask which of these genes will become more numerous in the gene pool. Superficially it might seem that the gene for laying four eggs is bound to have an advantage over the genes for laying three or two. A moment's reflection shows that this simple 'more means better' argument cannot be true, however. It leads to the expectation that five eggs should be better than four, ten better still, 100 even better, and infinity best of all. In other words it leads logically to an absurdity. Obviously there are *costs* as well as benefits in laying a large number of eggs. Increased bearing is bound to be paid for in less efficient caring. Lack's essential point is that for any given species, in any given environmental situation, there must be an optimal clutch size. Where he differs from Wynne-Edwards is in his answer to the question 'optimal from whose point of view?'. Wynne-Edwards would say the important optimum, to which all individuals should aspire, is the optimum for the group as a whole. Lack would say each selfish individual chooses the clutch size which maximizes the number of children she rears. If three is the optimum clutch size for swifts, what this means, for Lack, is that any individual who tries to rear four will probably end up with fewer children than rival, more cautious, individuals who only try to rear three. The obvious reason for this would be that the food is so thinly spread between the four babies that few of them survive to adulthood. This would be true both of the original allocation of yolk to the four eggs, and of the food given to the babies after hatching. According to Lack, therefore, individuals regulate their clutch size for reasons which are anything but altruistic. They are not practising birth-control in order to avoid over-exploiting the group's resources. They are practising birth-control in order to maximize the number of surviving children they actually have, an aim which is the very opposite of that which we normally associate with birth-control.

Rearing baby birds is a costly business. The mother has to invest a large quantity of food and energy in manufacturing eggs. Possibly with her mate's help, she invests a large effort in building a nest to hold her eggs and protect them. Parents spend weeks patiently sitting on the eggs. Then, when the babies hatch out, the parents work themselves nearly to death fetching food for them, more or less non-stop without resting. As we have already seen, a parent great tit brings an average of one item of food to the nest every 30 seconds of daylight. Mammals such as ourselves do it in a slightly different way, but the basic idea of reproduction being a costly affair, especially for the mother, is no less true. It is obvious that if a parent tries to spread her limited resources of food and effort among too many children, she will end up rearing fewer than if she had set out with more modest ambitions. She has to strike a balance between bearing and caring. The total amount of food and other resources which an individual female, or a mated pair, can muster is the limiting factor determining the number of children they can rear. Natural selection, according to the Lack theory, adjusts initial clutch size (litter size etc.) so as to take maximum advantage of these limited resources.

Individuals who have too many children are penalized, not because the whole population goes extinct, but simply because fewer of their children survive. Genes for having too many children are just not passed on to the next generation in large numbers, because few of the children bearing these genes reach adulthood. What has happened in modern civilized man is that family sizes are no longer limited by the finite resources which the individual parents can provide. If a husband and wife have more children than they can feed, the state, which means the rest of the population, simply steps in and keeps the surplus children alive and healthy. There is, in fact, nothing to stop a couple with no material resources at all having and rearing precisely as many children as the woman can physically bear. But the welfare state is a very unnatural thing. In nature, parents who have more children than they can support do not have many grandchildren, and their genes are not passed on to future generations. There is no *need* for altruistic restraint in the birth-rate, because there is no welfare state in nature. Any gene for over-indulgence is promptly punished: the children containing that gene starve. Since we humans do not want to return to the old selfish ways where we let the children of too-large families starve to death, we have abolished the family as a unit of economic self-sufficiency, and substituted the state. But the privilege of guaranteed support for children should not be abused.

Contraception is sometimes attacked as 'unnatural'. So it is, very unnatural. The trouble is, so is the welfare state. I think that most of us believe the welfare state is highly desirable. But you cannot have an unnatural welfare state, unless you also have unnatural birth-control, otherwise the end result will be misery even greater than that which obtains in nature. The welfare state is perhaps the greatest altruistic system the animal kingdom has ever known. But any altruistic system is inherently unstable, because it is open to abuse by selfish individuals, ready to exploit it. Individual humans who have more children than they are capable of rearing are probably too ignorant in most cases to be accused of conscious malevolent exploitation. Powerful institutions and leaders who deliberately encourage them to do so seem to me less free from suspicion.

Returning to wild animals, the Lack clutch-size argument can be generalized to all the other examples Wynne-Edwards uses: territorial behaviour, dominance hierarchies, and so on. Take, for instance, the red grouse that he and his colleagues have worked on. These birds eat heather, and they parcel out the moors in territories containing apparently more food than the territory owners actually need. Early in the season they fight over territories, but after a while the losers seem to accept that they have failed, and do not fight any more. They become outcasts who never get territories, and by the end of the season they have mostly starved to death. Only territory owners breed. That non-territory owners are physically capable of breeding is shown by the fact that if a territory owner is shot his place is promptly filled by one of the former outcasts, who then breeds. Wynne-Edwards's interpretation of this extreme territorial behaviour is, as we have seen, that the outcasts 'accept' that they have failed to gain a ticket or licence to breed; they do not try to breed.

On the face of it, this seems an awkward example for the selfish gene theory to explain. Why don't the outcasts try, try, and try again to oust a territory holder, until they drop from exhaustion? They would seem to have nothing to lose. But wait, perhaps they do have something to lose. We have already seen that if a terri-tory-holder should happen to die, an outcast has a chance of taking his place, and therefore of breeding. If the odds of an outcast's succeeding to a territory in this way are greater than the odds of his gaining one by fighting, then it may pay him, as a selfish individual, to wait in the hope that somebody will die, rather than squander what little energy he has in futile fighting. For Wynne-Edwards, the role of the outcasts in the welfare of the group is to wait in the wings as understudies, ready to step into the shoes of any territory holder who dies on the main stage of group reproduction. We can now see that this may also be their best strategy purely as selfish individuals. As we saw in Chapter 4, we can regard animals as gamblers. The best strategy for a gambler may sometimes be a wait-and-hope strategy, rather than a bull-at-a-gate strategy.

Similarly, the many other examples where animals appear to 'accept' non-reproductive status passively can be explained quite easily by the selfish gene theory. The general form of the explana-tion is always the same: the individual's best bet is to restrain himself for the moment, in the hope of better chances in the future. A seal who leaves the harem-holders unmolested is not doing it for the good of the group. He is biding his time, waiting for a more propitious moment. Even if the moment never comes and he ends up without descendants, the gamble *might* have paid off, though, with hindsight we can see that for him it did not.

And when lemmings flood in their millions away from the centre of a population explosion, they are not doing it in order to reduce the density of the area they leave behind! They are seeking, every selfish one of them, a less crowded place in which to live. The fact that any particular one may fail to find it, and dies, is something we can see with hindsight. It does not alter the likelihood that to stay behind would have been an even worse gamble.

It is a well-documented fact that overcrowding sometimes reduces birth-rates. This is sometimes taken to be evidence for Wynne-Edwards's theory. It is nothing of the kind. It is compatible with his theory, and it is also just as compatible with the selfish gene theory. For example, in one experiment mice were put in an outdoor enclosure with plenty of food, and allowed to breed freely. The population grew up to a point, then levelled off. The reason for the levelling-off turned out to be that the females became less fertile as a consequence of over-crowding: they had fewer babies. This kind of effect has often been reported. Its immediate cause is often called 'stress', although giving it a name like that does not of itself help to explain it. In any case, whatever its immediate cause may be, we still have to ask about its ultimate, or evolutionary explanation. Why does natural selection favour females who reduce their birth-rate when their population is over-crowded?

Wynne-Edwards's answer is clear. Group selection favours groups in which the females measure the population and adjust their birth-rates so that food supplies are not over-exploited. In the condition of the experiment, it so happened that food was never going to be scarce, but the mice could not be expected to realize that. They are programmed for life in the wild, and it is likely that in natural conditions over-crowding is a reliable indicator of future famine.

What does the selfish gene theory say? Almost exactly the same thing, but with one crucial difference. You will remember that, according to Lack, animals will tend to have the optimum number of children from their own selfish point of view. If they *bear* too few or too many, they will end up *rearing* fewer than they would have if they had hit on just the right number. Now, 'just the right number' is likely to be a smaller number in a year when the population is over-crowded than in a year when the population is sparse. We have already agreed that over-crowding is likely to foreshadow famine. Obviously, if a female is presented with reliable evidence that a famine is to be expected, it is in her own selfish interests to reduce her own birth-rate. Rivals who do not respond to the warning signs in this way will end up rearing fewer babies, even if they actually bear more. We therefore end up with almost exactly the same conclusion as Wynne-Edwards, but we get there by an entirely different type of evolutionary reasoning.

The selfish gene theory has no trouble even with 'epideictic displays'. You will remember that Wynne-Edwards hypothesized that animals deliberately display together in large crowds in order to make it easy for all the individuals to conduct a census, and regulate their birth-rates accordingly. There is no direct evidence that any aggregations are in fact epideictic, but just suppose some such evidence were found. Would the selfish gene theory be embarrassed? Not a bit.

Starlings roost together in huge numbers. Suppose it were shown, not only that over-crowding in winter reduced fertility in the following spring, but that this was directly due to the birds' listening to each other's calls. It might be demonstrated experimentally that individuals exposed to a tape-recording of a dense and very loud starling roost laid fewer eggs than individuals exposed to a recording of a quieter, less dense, roost. By definition, this would indicate that the calls of starlings constituted an epideictic display. The selfish gene theory would explain it in much the same way as it handled the case of the mice.

Again, we start from the assumption that genes for having a larger family than you can support are automatically penalized, and become less numerous in the gene pool. The task of an efficient egg-layer is one of predicting what is going to be the optimum clutch size for her, as a selfish individual, in the coming breeding season. You will remember from Chapter 4 the special sense in which we are using the word prediction. Now how can a female bird predict her optimum clutch size? What variables should influence her prediction? It may be that many species make a fixed prediction, which does not change from year to year. Thus on average the optimum clutch size for a gannet is one. It is possible that in particular bumper years for fish the true optimum for an individual might temporarily rise to two eggs. If there is no way for gannets to know in advance whether a particular year is going to be a bumper one, we cannot expect individual females to take the risk of wasting their resources on two eggs, when this would damage their reproductive success in an average year.

But there may be other species, perhaps starlings, in which it is in principle possible to predict in winter whether the following spring is going to yield a good crop of some particular food resource. Country people have numerous old sayings suggesting that such clues as the abundance of holly berries may be good predictors of the weather in the coming spring. Whether any particular old wives' tale is accurate or not, it remains logically possible that there are such clues, and that a good prophet could in theory adjust her clutch size from year to year to her own advantage. Holly berries may be reliable predictors or they may not but, as in the case of the mice, it does seem quite likely that population density would be a good predictor. A female starling can in principle know that, when she comes to feed her babies in the coming spring, she will be competing for food with rivals of the same species. If she can somehow estimate the local density of her own species in winter, this could provide her with a powerful means of predicting how difficult it is going to be to get food for babies next spring. If she found the winter population to be particularly high, her prudent policy, from her own selfish point of view, might well be to lay relatively few eggs: her estimate of her own optimum clutch size would have been reduced.

Now the moment it becomes true that individuals are reducing their clutch size on the basis of their estimate of population density, it will immediately be to the advantage of each selfish individual to pretend to rivals that the population is large, whether it really is or not. If starlings are estimating population size by the volume of noise in a winter roost, it would pay each individual to shout as loudly as possible, in order to sound more like two starlings than one. This idea of animals pretending to be several animals at once has been suggested in another context by J. R. Krebs, and is named the *Beau Geste Effect* after the novel in which a similar tactic was used by a unit of the French Foreign Legion. The idea in our case is to try to induce neighbouring starlings to reduce *their* clutch size to a level lower than the true optimum. If you are a starling who succeeds in doing this, it is to your selfish advantage, since you are reducing the numbers of individuals who do not bear your genes. I therefore conclude that Wynne-Edwards's idea of epideictic displays may actually be a good idea: he may have been right all along, but for the wrong reasons. More generally, the Lack type of hypothesis is powerful enough to account, in selfish gene terms, for all evidence that might seem to support the group-selection theory, should any such evidence turn up.

Our conclusion from this chapter is that individual parents practise family planning, but in the sense that they optimize their birth-rates rather than restrict them for public good. They try to maximize the number of surviving children that they have, and this means having neither too many babies nor too few. Genes which make an individual have too many babies tend not to persist in the gene pool, because children containing such genes tend not to survive to adulthood.

So much, then, for quantitative considerations of family size. We now come on to conflicts of interest within families. Will it always pay a mother to treat all her children equally, or might she have favourites? Should the family function as a single cooperating whole, or are we to expect selfishness and deception even within the family? Will all members of a family be working towards the same optimum, or will they 'disagree' about what the optimum is? These are the questions we try to answer in the next chapter. The related question of whether there may be conflict of interest between mates, we postpone until Chapter 9.

8. Battle of the generations

LET us begin by tackling the first of the questions posed at the end of the last chapter. Should a mother have favourites, or should she be equally altruistic towards all her children? At the risk of being boring, I must yet again throw in my customary warning. The word 'favourite' carries no subjective connotations, and the word 'should' no moral ones. I am treating a mother as a machine programmed to do everything in its power to propagate copies of the genes which ride inside it. Since you and I are humans who know what it is like to have conscious purposes, it is convenient for me to use the language of purpose as a metaphor in explaining the behaviour of survival machines.

In practice, what would it mean to say a mother had a favourite child? It would mean she would invest her resources unequally among her children. The resources which a mother has available to invest consist of a variety of things. Food is the obvious one, together with the effort expended in gathering food, since this in itself costs the mother something. Risk undergone in protecting young from predators is another resource which the mother can 'spend' or refuse to spend. Energy and time devoted to nest or home maintenance, protection from the elements, and, in some species, time spent in teaching children, are valuable resources which a parent can allocate to children, equally or unequally as she 'chooses'.

It is difficult to think of a common currency in which to measure all these resources which a parent can invest. Just as human societies use money as a universally convertible currency which can be translated into food or land or labouring time, so we require a currency in which to measure resources which an individual survival machine may invest in another individual's life, in particular a child's life. A measure of energy such as the calorie is tempting, and some ecologists have devoted themselves to the accounting of energy costs in nature. This is inadequate though, because it is only loosely convertible into the currency which really matters, the 'gold-standard' of evolution, gene survival. R. L. Trivers, in 1972, neatly solved the problem with his concept of *Parental Investment* (although, reading between the close-packed lines, one feels that Sir Ronald Fisher, the greatest biologist of the twentieth century, meant much the same thing in 1930 by his 'parental expenditure').

Parental Investment (P.I.) is defined as 'any investment by the parent in an individual offspring that increases the offspring's chance of surviving (and hence reproductive success) at the cost of the parent's ability to invest in other offspring.' The beauty of Trivers's parental investment is that it is measured in units very close to the units which really matter. When a child uses up some of its mother's milk, the amount of milk consumed is measured not in pints, not in calories, but in units of detriment to other children of the same mother. For instance, if a mother has two babies, X and Y, and X drinks one pint of milk, a major part of the P.I. which this pint represents is measured in units of increased probability that Y will die because he did not drink that pint. P.I. is measured in units of decrease in life expectancy of other children, born or yet to be born.

Parental investment is not quite an ideal measure, because it overemphasizes the importance of parentage, as against other genetic relationships. Ideally we should use a generalized *altruism investment* measure. Individual A may be said to invest in individual B, when A increases B's chance of surviving, at the cost of A's ability to invest in other individuals including herself, all costs being weighted by the appropriate relatedness. Thus a parent's investment in any one child should ideally be measured in terms of detriment to life expectancy not only of other children, but also of nephews, nieces, herself, etc. In many respects, however, this is just a quibble, and Trivers's measure is well worth using in practice.

Now any particular adult individual has, in her whole lifetime, a certain total quantity of P.I. available to invest in children (and other relatives and in herself, but for simplicity we consider only children). This represents the sum of all the food she can gather or manufacture in a lifetime of work, all the risks she is prepared to take, and all the energy and effort that she is able to put into the welfare of children. How should a young female, setting out on her adult life, invest her life's resources? What would be a wise investment policy for her to follow? We have already seen from the Lack theory that she should not spread her investment too thinly among too many children. That way she will lose too many genes: she won't have enough grandchildren. On the other hand, she must not devote all her investment to too few children— spoilt brats. She may virtually guarantee herself *some* grandchildren, but rivals who invest in the optimum number of children will end up with more grandchildren. So much for even-handed investment policies. Our present interest is in whether it could ever pay a mother to invest unequally among her children, i.e. in whether she should have favourites.

The answer is that there is no genetic reason for a mother to have favourites. Her relatedness to all her children is the same, $\frac{1}{2}$. Her optimal strategy is to invest *equally* in the largest number of children that she can rear to the age when they have children of their own. But, as we have already seen, some individuals are better life insurance risks than others. An under-sized runt bears just as many of his mother's genes as his more thriving litter mates. But his life expectation is less. Another way to put this is that he *needs* more than his fair share of parental investment, just to end up equal to his brothers. Depending on the circumstances, it may pay a mother to refuse to feed a runt, and allocate all of his share of her parental investment to his brothers and sisters. Indeed it may pay her to feed him to his brothers and sisters, or to eat him herself, and use him to make milk. Mother pigs do sometimes devour their young, but I do not know whether they pick especially on runts.

Runts constitute a particular example. We can make some more general predictions about how a mother's tendency to invest in a child might be affected by his age. If she has a straight choice between saving the life of one child or saving the life of another, and if the one she does not save is bound to die, she should prefer the older one. This is because she stands to lose a higher proportion of her life's parental investment if he dies than if his little brother dies. Perhaps a better way to put this is that if she saves the little brother she will still have to invest some costly resources in him just to get him up to the age of the big brother.

On the other hand, if the choice is not such a stark life or death choice, her best bet might be to prefer the younger one. For instance, suppose her dilemma is whether to give a particular morsel of food to a little child or a big one. The big one is likely to be more capable of finding his own food unaided. Therefore if she stopped feeding him he would not necessarily die. On the other hand, the little one who is too young to find food for himself would be more likely to die if his mother gave the food to his big brother. Now, even though the mother would prefer the little brother to die rather than the big brother, she may still give the food to the little one, because the big one is unlikely to die anyway. This is why mammal mothers wean their children, rather than going on feeding them indefinitely throughout their lives. There comes a time in the life of a child when it pays the mother to divert investment from him into future children. When this moment comes, she will want to wean him. A mother who had some way of knowing that she had had her last child might be expected to continue to invest all her resources in him for the rest of her life, and perhaps suckle him well into adulthood. Nevertheless, she should 'weigh up' whether it would not pay her more to invest in grandchildren or nephews and nieces, since although these are half as closely related to her as her own children, their capacity to benefit from her investment may be more than double that of one of her own children.

This seems a good moment to mention the puzzling phenomenon known as the menopause, the rather abrupt termination of a human female's reproductive fertility in middle age. This may not have occurred too commonly in our wild ancestors, since not many women would have lived that long anyway. But still, the difference between the abrupt change of life in women and the gradual fading out of fertility in men suggests that there is something genetically 'deliberate' about the menopause—that it is an 'adaptation'. It is rather difficult to explain. At first sight we might expect that a woman should go on having children until she dropped, even if advancing years made it progressively less likely that any individual child would survive. Surely it would seem always worth trying? But we must remember that she is also related to her grandchildren, though half as closely.

For various reasons, perhaps connected with the Medawar

theory of ageing (page 42), women in the natural state became gradually less efficient at bringing up children as they got older. Therefore the life expectancy of a child of an old mother was less than that of a child of a young mother. This means that, if a woman had a child and a grandchild born on the same day, the grandchild could expect to live longer than the child. When a woman reached the age where the average chance of each child reaching adulthood was just less than half the chance of each grandchild of the same age reaching adulthood, any gene for investing in grandchildren in preference to children would tend to prosper. Such a gene is carried by only one in four grandchildren, whereas the rival gene is carried by one in two children, but the greater expectation of life of the grandchildren outweighs this, and the 'grandchild altruism' gene prevails in the gene pool. A woman could not invest fully in her grandchildren if she went on having children of her own. Therefore genes for becoming reproductively infertile in middle age became more numerous, since they were carried in the bodies of grandchildren whose survival was assisted by grandmotherly altruism.

This is a possible explanation of the evolution of the menopause in females. The reason why the fertility of males tails off gradually rather than abruptly is probably that males do not invest so much as females in each individual child anyway. Provided he can sire children by young women, it will always pay even a very old man to invest in children rather than in grandchildren.

So far, in this chapter and in the last, we have seen everything from the parent's point of view, largely the mother's. We have asked whether parents can be expected to have favourites, and in general what is the best investment policy for a parent. But perhaps each child can influence how much his parents invest in him as against his brothers and sisters. Even if parents do not 'want' to show favouritism among their children, could it be that children grab favoured treatment for themselves? Would it pay them to do so? More strictly, would genes for selfish grabbing among children become more numerous in the gene pool than rival genes for accepting no more than one's fair share? This matter has been brilliantly analysed by Trivers, in a paper of 1974 called *Parent–Offspring Conflict*.

A mother is equally related to all her children, born and to be born. On genetic grounds alone she should have no favourites, as we have seen. If she does show favouritism it should be based on differences in expectation of life, depending on age and other things. The mother, like any individual, is twice as closely 'related' to herself as she is to any of her children. Other things being equal, this means that she should invest most of her resources selfishly in herself, but other things are not equal. She can do her genes more good by investing a fair proportion of her resources in her children. This is because these are younger and more helpless than she is, and they can therefore benefit more from each unit of investment than she can herself. Genes for investing in more helpless individuals in preference to oneself can prevail in the gene pool, even though the beneficiaries may share only a proportion of one's genes. This is why animals show parental altruism, and indeed why they show any kind of kin-selected altruism.

Now look at it from the point of view of a particular child. He is just as closely related to each of his brothers and sisters as his mother is to them. The relatedness is $\frac{1}{2}$ in all cases. Therefore he 'wants' his mother to invest some of her resources in his brothers and sisters. Genetically speaking, he is just as altruistically disposed to them as his mother is. But again, he is twice as closely related to himself as he is to any brother or sister, and this will dispose him to want his mother to invest in him more than in any particular brother or sister, other things being equal. In this case other things might indeed be equal. If you and your brother are the same age, and both are in a position to benefit equally from a pint of mother's milk, you 'should' try to grab more than your fair share, and he should try to grab more than his fair share. Have you ever heard a litter of piglets squealing to be first on the scene when the mother sow lies down to feed them? Or little boys fighting over the last slice of cake? Selfish greed seems to characterize much of child behaviour.

But there is more to it than this. If I am competing with my brother for a morsel of food, and if he is much younger than me so that he could benefit from the food more than I could, it might pay my genes to let him have it. An elder brother may have exactly the same grounds for altruism as a parent: in both cases, as we have seen, the relatedness is $\frac{1}{2}$, and in both cases the younger individual can make better use of the resource than the elder. If I possess a gene for giving up food, there is a 50 per cent chance that my baby brother contains the same gene. Although the gene has double the chance of being in my own body—100 per cent, it *is* in my body—my need of the food may be less than half as urgent. In general, a child 'should' grab more than his share of parental investment, but only up to a point. Up to what point? Up to the point where the resulting net cost to his brothers and sisters, born and potentially to be born, is just double the benefit of the grabbing to himself.

Consider the question of when weaning should take place. A mother wants to stop suckling her present child so that she can prepare for the next one. The present child, on the other hand, does not want to be weaned yet, because milk is a convenient, trouble-free source of food, and he does not want to have to go out and work for his living. To be more exact, he does want eventually to go out and work for his living, but only when he can do his genes more good by leaving his mother free to rear his little brothers and sisters, than by staying behind himself. The older a child is, the less relative benefit does he derive from each pint of milk. This is because he is bigger, and a pint of milk is therefore a smaller proportion of his requirement, and also he is becoming more capable of fending for himself if he is forced to. Therefore when an old child drinks a pint which could have been invested in a younger child, he is taking relatively more parental invest-ment for himself than when a young child drinks a pint. As a child grows older, there will come a moment when it would pay his mother to stop feeding him, and invest in a new child instead. Somewhat later there will come a time when the old child too would benefit his genes most by weaning himself. This is the moment when a pint of milk can do more good to the copies of his genes which *may be* present in his brothers and sisters than it can to the genes which *are* present in himself.

The disagreement between mother and child is not an absolute one, but a quantitative one, in this case a disagreement over timing. The mother wants to go on suckling her present child up to the moment when investment in him reaches his 'fair' share, taking into account his expectation of life and how much she has already invested in him. Up to this point there is no disagreement. Similarly, both mother and child agree in not wanting him to go on sucking after the point when the cost to future children is more than double the benefit to him. But there is disagreement between mother and child during the intermediate period, the period when the child is getting more than his share as the mother sees it, but when the cost to other children is still less than double the benefit to him.

Weaning time is just one example of a matter of dispute between mother and child. It could also be regarded as a dispute between one individual and all his future unborn brothers and sisters, with the mother taking the part of her future unborn children. More directly there may be competition between contemporary rivals for her investment, between litter mates or nest mates. Here, once again, the mother will normally be anxious to see fair play.

Many baby birds are fed in the nest by their parents. They all gape and scream, and the parent drops a worm or other morsel in the open mouth of one of them. The loudness with which each baby screams is, ideally, proportional to how hungry he is. Therefore, if the parent always gives the food to the loudest screamer, they should all tend to get their fair share, since when one has had enough he will not scream so loudly. At least that is what would happen in the best of all possible worlds, if individuals did not cheat. But in the light of our selfish gene concept we must expect that individuals *will* cheat, *will* tell lies about how hungry they are. This will escalate, apparently rather pointlessly because it might seem that if they are all lying by screaming too loudly, this level of loudness will become the norm, and will cease, in effect, to be a lie. However, it cannot de-escalate, because any individual who takes the first step in decreasing the loudness of his scream will be penalized by being fed less, and is more likely to starve. Baby bird screams do not become infinitely loud, because of other considerations. For example, loud screams tend to attract predators, and they use up energy.

Sometimes, as we have seen, one member of a litter is a runt, much smaller than the rest. He is unable to fight for food as strongly as the rest, and runts often die. We have considered the conditions under which it would actually pay a mother to let a runt die. We might suppose intuitively that the runt himself should go on struggling to the last, but the theory does not necessarily predict this. As soon as a runt becomes so small and weak that his expectation of life is reduced to the point where benefit to him due to parental investment is less than twice the benefit which the same investment could potentially confer on the other babies, the runt should die gracefully and willingly. He can benefit his genes most by doing so. That is to say, a gene which gives the instruction 'Body, if you are very much smaller than your litter-mates, give up the struggle and die', could be successful in the gene pool, because it has a 50 per cent chance of being in the body of each brother and sister saved, and its chances of surviving in the body of the runt are very small anyway. There should be a point of no return in the career of a runt. Before he reaches this point he should go on struggling. As soon as he reaches it he should give up, and preferably let himself be eaten by his litter-mates or his parents.

I did not mention it when we were discussing Lack's theory of clutch size, but the following is a reasonable strategy for a parent who is undecided as to what is her optimum clutch size for the current year. She might lay one more egg than she actually 'thinks' is likely to be the true optimum. Then, if the year's food crop should turn out to be a better one than expected, she will rear the extra child. If not, she can cut her losses. By being careful always to feed the young in the same order, say in order of size, she sees to it that one, perhaps a runt, quickly dies, and not too much food is wasted on him, beyond the initial investment of egg yolk or equivalent. From the mother's point of view, this may be the explanation of the runt phenomenon. He represents the hedging of the mother's bets. This has been observed in many birds.

Using our metaphor of the individual animal as a survival machine behaving as if it had the 'purpose' of preserving its genes, we can talk about a conflict between parents and young, a battle of the generations. The battle is a subtle one, and no holds are barred on either side. A child will lose no opportunity of cheating. It will pretend to be hungrier than it is, perhaps younger than it is, more in danger than it really is. It is too small and weak to bully its parents physically, but it uses every psychological weapon at its disposal: lying, cheating, deceiving, exploiting, right up to the point where it starts to penalize its relatives more than its genetic relatedness to them should allow. Parents, on the other hand, must be alert to cheating and deceiving, and must try not to be fooled by it. This might seem an easy task. If the parent knows that its child is likely to lie about how hungry it is, it might employ the tactic of feeding it a fixed amount and no more, even though the child goes on screaming. One trouble with this is that the child may not have been lying, and if it dies as a result of not being fed the parent would have lost some of its precious genes. Wild birds can die after being starved for only a few hours.

A. Zahavi has suggested a particularly diabolical form of child blackmail: the child screams in such a way as to attract predators deliberately to the nest. The child is 'saying' 'Fox, fox, come and get me.' The only way the parent can stop it screaming is to feed it. So the child gains more than its fair share of food, but at a cost of some risk to itself. The principle of this ruthless tactic is the same as that of the hijacker threatening to blow up an aeroplane, with himself on board, unless he is given a ransom. I am sceptical about whether it could ever be favoured in evolution, not because it is too ruthless, but because I doubt if it could ever pay the blackmailing baby. He has too much to lose if a predator really came. This is clear for an only child, which is the case Zahavi himself considers. No matter how much his mother may already have invested in him, he should still value his own life more than his mother values it, since she has only half of his genes. Moreover, the tactic would not pay even if the blackmailer was one of a clutch of vulnerable babies, all in the nest together, since the blackmailer has a 50 per cent genetic 'stake' in each of his endangered brothers and sisters, as well as a 100 per cent stake in himself. I suppose the theory might conceivably work if the predominant predator had the habit of only taking the largest nestling from a nest. Then it might pay a smaller one to use the threat of summoning a predator, since it would not be greatly endangering itself. This is analogous to holding a pistol to your brother's head rather than threatening to blow yourself up.

More plausibly, the blackmail tactic might pay a baby cuckoo. As is well known, cuckoo females lay one egg in each of several 'foster' nests, and then leave the unwitting foster-parents, of a quite different species, to rear the cuckoo young. Therefore a baby cuckoo has no genetic stake in his foster brothers and sisters. (Some species of baby cuckoo will not have any foster brothers and sisters, for a sinister reason which we shall come to. For the moment I assume we are dealing with one of those species in which foster brothers and sisters co-exist alongside the baby cuckoo.) If a baby cuckoo screamed loudly enough to attract predators, it would have a lot to lose—its life—but the foster mother would have even more to lose, perhaps four of her young. It could therefore pay her to feed it more than its share, and the advantage of this to the cuckoo might outweigh the risk.

This is one of those occasions when it would be wise to translate back into respectable gene language, just to reassure ourselves that we have not become too carried away with subjective metaphors. What does it really mean to set up the hypothesis that baby cuckoos 'blackmail' their foster parents by screaming 'Predator, predator, come and get me and all my little brothers and sisters'? In gene terms it means the following.

Cuckoo genes for screaming loudly became more numerous in the cuckoo gene pool because the loud screams increased the probability that the foster parents would feed the baby cuckoos. The reason the foster parents responded to the screams in this way was that genes for responding to the screams had spread through the gene pool of the foster-species. The reason these genes spread was that individual foster parents who did not feed the cuckoos extra food, reared fewer of their own children—fewer than rival parents who did feed their cuckoos extra. This was because predators were attracted to the nest by the cuckoo cries. Although cuckoo genes for not screaming were less likely to end up in the bellies of predators than screaming genes, the non-screaming cuckoos paid the greater penalty of not being fed extra rations. Therefore the screaming genes spread through the cuckoo gene pool.

A similar chain of genetic reasoning, following the more subjective argument given above, would show that although such a blackmailing gene could conceivably spread through a cuckoo gene pool, it is unlikely to spread through the gene pool of an ordinary species, at least not for the specific reason that it attracted predators. Of course, in an ordinary species there could be other reasons for screaming genes to spread, as we have already seen, and these would *incidentally* have the effect of occasionally attracting predators. But here the selective influence of predation would be, if anything, in the direction of making the cries quieter. In the hypothetical case of the cuckoos, the net influence of predators, paradoxical as it sounds at first, could be to make the cries louder.

There is no evidence, one way or the other, on whether cuckoos, and other birds of similar 'brood-parasitic' habit, actually employ the blackmail tactic. But they certainly do not lack ruthlessness. For instance, there are honeyguides who, like cuckoos, lay their eggs in the nests of other species. The baby honeyguide is equipped with a sharp, hooked beak. As soon as he hatches out, while he is still blind, naked, and otherwise helpless, he scythes and slashes his foster brothers and sisters to death: dead brothers do not compete for food! The familiar British cuckoo achieves the same result in a slightly different way. It has a short incubation-time, and so the baby cuckoo manages to hatch out before its foster brothers and sisters. As soon as it hatches, blindly and mechanically, but with devastating effectiveness, it throws the other eggs out of the nest. It gets underneath an egg, fitting it into a hollow in its back. Then it slowly backs up the side of the nest, balancing the egg between its wing-stubs, and topples the egg out on to the ground. It does the same with all the other eggs, until it has the nest, and therefore the attention of its foster parents, entirely to itself.

One of the most remarkable facts I have learned in the past year was reported from Spain by F. Alvarez, L. Arias de Reyna, and H. Segura. They were investigating the ability of potential foster parents—potential victims of cuckoos—to detect intruders, cuckoo eggs or chicks. In the course of their experiments they had occasion to introduce into magpie nests the eggs and chicks of cuckoos, and, for comparison, eggs and chicks of other species such as swallows. On one occasion they introduced a baby swallow into a magpie's nest. The next day they noticed one of the magpie eggs lying on the ground under the nest. It had not broken, so they picked it up, replaced it, and watched. What they saw is utterly remarkable. The baby swallow, behaving exactly as if it was a baby cuckoo, threw the egg out. They replaced the egg again, and exactly the same thing happened. The baby swallow used the cuckoo method of balancing the egg on its back between its wing-stubs, and walking backwards up the side of the nest until the egg toppled out.

Perhaps wisely, Alvarez and his colleagues made no attempt to explain their astonishing observation. How could such behaviour evolve in the swallow gene pool? It must correspond to something in the normal life of a swallow. Baby swallows are not accustomed to finding themselves in magpie nests. They are never normally found in any nest except their own. Could the behaviour represent an evolved anti-cuckoo adaptation? Has natural selection been favouring a policy of counter-attack in the swallow gene pool, genes for hitting the cuckoo with his own weapons? It seems to be a fact that swallows' nests are not normally parasitized by cuckoos. Perhaps this is why. According to this theory, the magpie eggs of the experiment would be incidentally getting the same treatment, perhaps because, like cuckoo eggs, they are bigger than swallow eggs. But if baby swallows can tell the difference between a large egg and a normal swallow egg, surely the mother should be able to as well. In this case why is it not the mother who ejects the cuckoo egg, since it would be so much easier for her to do so than the baby? The same objection applies to the theory that the baby swallow's behaviour normally functions to remove addled eggs or other debris from the nest. Once again, this task could be—and is—performed better by the parent. The fact that the difficult and skilled egg-rejecting operation was seen to be performed by a weak and helpless baby swallow, whereas an adult swallow could surely do it much more easily, compels me to the conclusion that, from the parent's point of view, the baby is up to no good.

It seems to me just conceivable that the true explanation has nothing to do with cuckoos at all. The blood may chill at the thought, but could this be what baby swallows do to each other? Since the firstborn is going to compete with his yet unhatched brothers and sisters for parental investment, it could be to his advantage to begin his life by throwing out one of the other eggs.

The Lack theory of clutch size considered the optimum from the parent's point of view. If I am a mother swallow, the optimum clutch-size from my point of view is, say five. But if I am a baby swallow, the optimum clutch size as I see it may well be a smaller number, provided I am one of them! The parent has a certain amount of parental investment, which she 'wishes' to distribute even-handedly among five young. But each baby wants more than his allotted one fifth share. Unlike a cuckoo, he does not want all of it, because he is related to the other babies. But he does want more than one fifth. He can acquire a $\frac{1}{4}$ share simply by tipping out one egg; a $\frac{1}{3}$ share by tipping out another. Translating into gene language, a gene for fratricide could conceivably spread through the gene pool, because it has 100 per cent chance of being in the body of the fratricidal individual, and only a 50 per cent chance of being in the body of his victim.

The chief objection to this theory is that it is very difficult to believe that nobody would have seen this diabolical behaviour if it really occurred. I have no convincing explanation for this. There are different races of swallow in different parts of the world. It is known that the Spanish race differs from, for example, the British one, in certain respects. The Spanish race has not been subjected to the same degree of intensive observation as the British one, and I suppose it is just conceivable that fratricide occurs but has been overlooked.

My reason for suggesting such an improbable idea as the fratricide hypothesis here is that I want to make a general point. This is that the ruthless behaviour of a baby cuckoo is only an extreme case of what must go on in any family. Full brothers are more closely related to each other than a baby cuckoo is to its foster brothers, but the difference is only a matter of degree. Even if we cannot believe that outright fratricide could evolve, there must be numerous lesser examples of selfishness where the cost to the child, in the form of losses to his brothers and sisters, is outweighed, more than two to one, by the benefit to himself. In such cases, as in the example of weaning time, there is a real conflict of interests between parent and child.

Who is most likely to win this battle of the generations? R. D. Alexander has written an interesting paper in which he suggests that there is a general answer to this question. According to him the parent will always win. Now if this is the case, you have been wasting your time reading this chapter. If Alexander is right, much that is of interest follows. For instance, altruistic behaviour could evolve, not because of benefit to the genes of the individual himself, but solely because of benefit to his parents' genes. Parental manipulation, to use Alexander's term, becomes an alternative evolutionary cause of altruistic behaviour, independent of straightforward kin selection. It is therefore important that we examine Alexander's reasoning, and convince ourselves that we understand why he is wrong. This should really be done mathematically, but we are avoiding explicit use of mathematics in this book, and it is possible to give an intuitive idea of what is wrong with Alexander's thesis.

His fundamental genetic point is contained in the following abridged quotation. 'Suppose that a juvenile ... cause(s) an uneven distribution of parental benefits in its own favor, thereby reducing the mother's own overall reproduction. A gene which in this fashion improves an individual's fitness when it is a juvenile cannot fail to lower its fitness more when it is an adult, for such mutant genes will be present in an increased proportion of the mutant individual's offspring.' The fact that Alexander is considering a newly mutated gene is not fundamental to the argument. It is better to think of a rare gene inherited from one of the parents. 'Fitness' has the special technical meaning of reproductive success. What Alexander is basically saying is this. A gene which made a child grab more than his fair share when he was a child, at the expense of his parent's total reproductive output, might indeed increase his chances of surviving. But he would pay the penalty when he came to be a parent himself, because his own children would tend to inherit the same selfish gene, and this would reduce his overall reproductive success. He would be hoist with his own petard. Therefore the gene cannot succeed, and parents must always win the conflict.

Our suspicions should be immediately aroused by this argument, because it rests on the assumption of a genetic asymmetry which is not really there. Alexander is using the words 'parent' and 'offspring' as though there was a fundamental genetic difference between them. As we have seen, although there are *practical* differences between parent and child, for instance parents are older than children, and children come out of parents' bodies, there is really no fundamental *genetic* asymmetry. The relatedness is 50 per cent, whichever way round you look at it. To illustrate what I mean, I am going to repeat Alexander's words, but with 'parent', 'juvenile' and other appropriate words reversed. 'Suppose that a *parent* has a gene which tends to cause an *even* distribution of parental benefits. A gene which in this fashion improves an individual's fitness when it is a *parent* could not fail to have lowered its fitness more when it was a *juvenile*.' We therefore reach the opposite conclusion to Alexander, namely that in any parent/offspring conflict, the child must win!

Obviously something is wrong here. Both arguments have been put too simply. The purpose of my reverse quotation is not to prove the opposite point to Alexander, but simply to show that you cannot argue in that kind of artificially asymmetrical way. Both Alexander's argument, and my reversal of it, erred through looking at things from the point of view of an *individual*—in Alexander's case, the parent, in my case, the child. I believe this kind of error is all too easy to make when we use the technical term 'fitness'. This is why I have avoided using the word in this book. There is really only one entity whose point of view matters in evolution, and that entity is the selfish gene. Genes in juvenile bodies will be selected for their ability to outsmart parental bodies; genes in parental bodies will be selected for their ability to outsmart the young. There is no paradox in the fact that the very same genes successively occupy a juvenile body and a parental body. Genes are selected for their ability to make the best use of the levers of power at their disposal: they will exploit their practical opportunities. When a gene is sitting in a juvenile body its practical opportunities will be different from when it is sitting in a parental body. Therefore its optimum policy will be different in the two stages in its body's life history. There is no reason to suppose, as Alexander does, that the later optimum policy should necessarily overrule the earlier.

There is another way of putting the argument against Alexander. He is tacitly assuming a false asymmetry between the parent/child relationship on the one hand, and the brother/sister relationship on the other. You will remember that, according to Trivers, the cost to a selfish child of grabbing more than his share, the reason why he only grabs up to a point, is the danger of loss of his brothers and sisters who each bear half his genes. But brothers and sisters are only a special case of relatives with a 50 per cent relatedness. The selfish child's own future children are no more and no less 'valuable' to him than his brothers and sisters. Therefore the total net cost of grabbing more than your fair share of resources should really be measured, not only in lost brothers and sisters, but also in lost future offspring due to their selfishness among themselves. Alexander's point about the disadvantage of juvenile selfishness spreading to your own children, thereby reducing your own long-term reproductive output, is well taken, but it simply means we must add this in to the cost side of the equation. An individual child will still do well to be selfish so long as the net benefit to him is at least double the net cost to close relatives. But 'close relatives' should be read as including, not just brothers and sisters, but future children of one's own as well. An individual should reckon his own welfare as twice as valuable as that of his brothers, which is the basic assumption Trivers makes. But he should also value himself twice as highly as one of his own future children. Alexander's conclusion that there is a built-in advantage on the parent's side in the conflict of interests is not correct.

In addition to his fundamental genetic point, Alexander also has more practical arguments, stemming from undeniable asymmetries in the parent/child relationship. The parent is the active partner, the one who actually does the work to get the food, etc., and is therefore in a position to call the tune. If the parent decides to withdraw its labour, there is not much that the child can do about it, since it is smaller, and cannot hit back. Therefore the parent is in a position to impose its will, regardless of what the child may want. This argument is not obviously wrong, since in this case the asymmetry which it postulates is a real one. Parents really are bigger, stronger and more worldly-wise than children. They seem to hold all the good cards. But the young have a few aces up their sleeves too. For example, it is important for a parent to know how hungry each of its children is, so that it can most efficiently dole out the food. It could of course ration the food exactly equally between them all, but in the best of all possible worlds this would be less efficient than a system of giving a little bit more to those who could genuinely use it best. A system whereby each child told the parent how hungry he was would be ideal for the parent, and, as we have seen, such a system seems to have evolved. But the young are in a strong position to lie, because they *know* exactly how hungry they are, while the parent can only *guess* whether they are telling the truth or not. It is almost impossible for a parent to detect a small lie, although it might see through a big one.

Then again, it is of advantage to a parent to know when a baby is happy, and it is a good thing for a baby to be able to tell its parents when it is happy. Signals like purring and smiling may have been selected because they enable parents to learn which of their actions are most beneficial to their children. The sight of her child smiling, or the sound of her kitten purring, is rewarding to a mother, in the same sense as food in the stomach is rewarding to a rat in a maze. But once it becomes true that a sweet smile or a loud purr are rewarding, the child is in a position to use the smile or the purr in order to manipulate the parent, and gain more than its fair share of parental investment.

There is, then, no general answer to the question of who is more likely to win the battle of the generations. What will finally emerge is a compromise between the ideal situation desired by the child and that desired by the parent. It is a battle comparable to that between cuckoo and foster parent, not such a fierce battle to be sure, for the enemies do have some genetic interests in common—they are only enemies up to a point, or during certain sensitive times. However, many of the tactics used by cuckoos, tactics of deception and exploitation, may be employed by a parent's own young, although the parent's own young will stop short of the total selfishness which is to be expected of a cuckoo.

This chapter, and the next in which we discuss conflict between mates, could seem horribly cynical, and might even be distressing to human parents, devoted as they are to their children, and to each other. Once again I must emphasize that I am not talking about conscious motives. Nobody is suggesting that children deliberately and consciously deceive their parents because of the selfish genes within them. And I must repeat that when I say something like 'A child should lose no opportunity of cheating . . . lying, deceiving, exploiting . . .', I am using the word 'should' in a special way. I am not advocating this kind of behaviour as moral or desirable. I am simply saying that natural selection will tend to favour children who do act in this way, and that therefore when we look at wild populations we may expect to see cheating and selfishness within families. The phrase 'the child should cheat' means that genes which tend to make children cheat have an advantage in the gene pool. If there is a human moral to be drawn, it is that we must *teach* our children altruism, for we cannot expect it to be part of their biological nature.

9. Battle of the sexes

IF there is conflict of interest between parents and children, who share 50 per cent of each others' genes, how much more severe must be the conflict between mates, who are not related to each other? All that they have in common is a 50 per cent genetic shareholding in the same children. Since father and mother are both interested in the welfare of different halves of the same children, there may be some advantage for both of them in cooperating with each other in rearing those children. If one parent can get away with investing less than his or her fair share of costly resources in each child, however, he will be better off, since he will have more to spend on other children by other sexual partners, and so propagate more of his genes. Each partner can therefore be thought of as trying to exploit the other, trying to force the other one to invest more. Ideally, what an individual would 'like' (I don't mean physically enjoy, although he might) would be to copulate with as many members of the opposite sex as possible, leaving the partner in each case to bring up the children. As we shall see, this state of affairs is achieved by the males of a number of species, but in other species the males are obliged to share an equal part of the burden of bringing up children. This view of sexual partnership, as a relationship of mutal mistrust and mutual exploitation, has been stressed especially by Trivers. It is a comparatively new one to ethologists. We had usually thought of sexual behaviour, copulation, and the courtship which precedes it, as essentially a cooperative venture undertaken for mutual benefit, or even for the good of the species!

Let us go right back to first principles, and inquire into the fundamental nature of maleness and femaleness. In Chapter 3 we discussed sexuality without stressing its basic asymmetry. We simply accepted that some animals are called male, and others female, without asking what these words really meant. But what is the essence of maleness? What, at bottom, defines a female? We

as mammals see the sexes defined by whole syndromes of characteristics—possession of a penis, bearing of the young, suckling by
means of special milk glands, certain chromosomal features, and
so on. These criteria for judging the sex of an individual are all
very well for mammals but, for animals and plants generally, they
are no more reliable than is the tendency to wear trousers as a
criterion for judging human sex. In frogs, for instance, neither
sex has a penis. Perhaps then the words male and female have no
general meaning. They are, after all, only words, and if we do not
find them helpful for describing frogs, we are quite at liberty to
abandon them. We could arbitrarily divide frogs into Sex 1 and
Sex 2 if we wished. However, there is one fundamental feature of
the sexes which can be used to label males as males, and females
as females, throughout animals and plants. This is that the sex
cells or 'gametes' of males are much smaller and more numerous
than the gametes of females. This is true whether we are dealing
with animals or plants. One group of individuals has large sex
cells, and it is convenient to use the word female for them. The
other group, which it is convenient to call male, has small sex
cells. The difference is especially pronounced in reptiles and in
birds, where a single egg cell is big enough and nutritious enough
to feed a developing baby for several weeks. Even in humans,
where the egg is microscopic, it is still many times larger than the
sperm. As we shall see, it is possible to interpret all the other
differences between the sexes as stemming from this one basic
difference.

In certain primitive organisms, for instance some fungi,
maleness and femaleness do not occur, although sexual reproduction of a kind does. In the system known as isogamy the
individuals are not distinguishable into two sexes. Anybody can
mate with anybody else. There are not two different sorts of
gametes—sperms and eggs—but all sex cells are the same, called
isogametes. New individuals are formed by the fusion of two
isogametes, each produced by meiotic division. If we have three
isogametes, A, B, and C, A could fuse with B or C, B could fuse
with A or C. The same is never true of normal sexual systems. If
A is a sperm and it can fuse with B or C, then B and C must be
eggs and B cannot fuse with C.

When two isogametes fuse, both contribute equal numbers of

genes to the new individual, and they also contribute equal amounts of food reserves. Sperms and eggs too contribute equal numbers of genes, but eggs contribute far more in the way of food reserves: indeed sperms make no contribution at all, and are simply concerned with transporting their genes as fast as possible to an egg. At the moment of conception, therefore, the father has invested less than his fair share (i.e. 50 per cent) of resources in the offspring. Since each sperm is so tiny, a male can afford to make many millions of them every day. This means he is potentially able to beget a very large number of children in a very short period of time, using different females. This is only possible because each new embryo is endowed with adequate food by the mother in each case. This therefore places a limit on the number of children a female can have, but the number of children a male can have is virtually unlimited. Female exploitation begins here.

Parker and others showed how this asymmetry might have evolved from an originally isogamous state of affairs. In the days when all sex cells were interchangeable and of roughly the same size, there would have been some which just happened to be slightly bigger than others. In some respects a big isogamete would have an advantage over an average-sized one, because it would get its embryo off to a good start by giving it a large initial food supply. There might therefore have been an evolutionary trend towards larger gametes. But there was a catch. The evolution of isogametes which were larger than was strictly necessary would have opened the door to selfish exploitation. Individuals who produced *smaller* than average gametes could cash in, provided they could ensure that their small gametes fused with extra-big ones. This could be achieved by making the small ones more mobile, and able to seek out large ones actively. The advantage to an individual of producing small, rapidly moving gametes would be that he could afford to make a larger number of gametes, and therefore could potentially have more children. Natural selection favoured the production of sex cells which were small, and which actively sought out big ones to fuse with. So we can think of two divergent sexual 'strategies' evolving. There was the large-investment or 'honest' strategy. This automatically opened the way for a small-investment exploitative or 'sneaky' strategy. Once the divergence between the two strategies had started, it would have continued in runaway fashion. Medium-sized intermediates would have been penalized, because they did not enjoy the advantages of either of the two more extreme stategies. The sneaky ones would have evolved smaller and smaller size, and faster mobility. The honest ones would have evolved larger and larger size, to compensate for the ever-smaller investment contributed by the sneaky ones, and they became immobile because they would always be actively chased by the sneaky ones anyway. Each honest one would 'prefer' to fuse with another honest one. But the selection pressure (page 38) to lock out sneaky ones would have been weaker than the pressure on sneaky ones to duck under the barrier: the sneaky ones had more to lose, and they therefore won the evolutionary battle. The honest ones became eggs, and the sneaky ones became sperms.

Males, then, seem to be pretty worthless fellows, and on simple 'good of the species' grounds, we might expect that males would become less numerous than females. Since one male can theoretically produce enough sperms to service a harem of 100 females we might suppose that females should outnumber males in animal populations by 100 to 1. Other ways of putting this are that the male is more 'expendable', and the female more 'valuable' to the species. Of course, looked at from the point of view of the species as a whole, this is perfectly true. To take an extreme example, in one study of elephant seals, 4 per cent of the males accounted for 88 per cent of all the copulations observed. In this case, and in many others, there is a large surplus of bachelor males who probably never get a chance to copulate in their whole lives. But these extra males live otherwise normal lives, and they eat up the population's food resources no less hungrily than other adults. From a 'good of the species' point of view this is horribly wasteful; the extra males might be regarded as social parasites. This is just one more example of the difficulties that the group selection theory gets into. The selfish gene theory, on the other hand, has no trouble in explaining the fact that the numbers of males and females tend to be equal, even when the males who actually reproduce may be a small fraction of the total number. The explanation was first offered by R. A. Fisher.

The problem of how many males and how many females are born is a special case of a problem in parental strategy. Just as we discussed the optimal family size for an individual parent trying to maximize her gene survival, we can also discuss the optimal sex ratio. Is it better to entrust your precious genes to sons or to daughters? Suppose a mother invested all her resources in sons, and therefore had none left to invest in daughters: would she on average contribute more to the gene pool of the future than a rival mother who invested in daughters? Do genes for preferring sons become more or less numerous than genes for preferring daughters? What Fisher showed is that under normal circumstances the optimal sex ratio is 50:50. In order to see why, we must first know a little bit about the mechanics of sex determination.

In mammals, sex is determined genetically as follows. All eggs are capable of developing into either a male or a female. It is the sperms which carry the sex-determining chromosomes. Half the sperms produced by a man are female-producing, or X-sperms, and half are male-producing, or Y-sperms. The two sorts of sperms look alike. They differ with respect to one chromosome only. A gene for making a father have nothing but daughters could achieve its object by making him manufacture nothing but X-sperms. A gene for making a mother have nothing but daughters could work by making her secrete a selective spermicide, or by making her abort male embryos. What we seek is something equivalent to an evolutionarily stable strategy (ESS), although here, even more than in the chapter on aggression, strategy is just a figure of speech. An individual cannot literally choose the sex of his children. But genes for tending to have children of one sex or the other are possible. If we suppose that such genes, favouring unequal sex ratios, exist, are any of them likely to become more numerous in the gene pool than their rival alleles, which favour an equal sex ratio?

Suppose that in the elephant seals mentioned above, a mutant gene arose which tended to make parents have mostly daughters. Since there is no shortage of males in the population, the daughters would have no trouble finding mates, and the daughter-manufacturing gene could spread. The sex ratio in the population might then start to shift towards a surplus of females. From the point of view of the good of the species, this would be all right, because just a few males are quite capable of providing all the sperms needed for even a huge surplus of females, as we have seen. Superficially, therefore, we might expect the daughter-producing gene to go on spreading until the sex ratio was so unbalanced that the few remaining males, working flat out, could just manage. But now, think what an enormous genetic advantage is enjoyed by those few parents who have sons. Anyone who invests in a son has a very good chance of being the grandparent of hundreds of seals. Those who are producing nothing but daughters are assured of a safe few grandchildren, but this is nothing compared to the glorious genetic possibilities which open up before anyone specializing in sons. Therefore genes for producing sons will tend to become more numerous, and the pendulum will swing back.

For simplicity I have talked in terms of a pendulum swing. In practice the pendulum would never have been allowed to swing that far in the direction of female domination, because the pressure to have sons would have started to push it back as soon as the sex ratio became unequal. The strategy of producing equal numbers of sons and daughters is an evolutionarily stable strategy, in the sense that any gene for departing from it makes a net loss.

I have told the story in terms of numbers of sons versus numbers of daughters. This is to make it simple, but strictly it should be worked out in terms of parental investment, meaning all the food and other resources which a parent has to offer, measured in the way discussed in the previous chapter. Parents should *invest* equally in sons and daughters. This usually means they should have numerically as many sons as they have daughters. But there could be unequal sex ratios which were evolutionarily stable, provided correspondingly unequal amounts of resources were invested in sons and daughters. In the case of the elephant seals, a policy of having three times as many daughters as sons, but of making each son a supermale by investing three times as much food and other resources in him, could be stable. By investing more food in a son and making him big and strong, a parent might increase his chances of winning the supreme prize of a harem. But this is a special case. Normally the amount invested in each son will roughly equal the amount invested in each daughter, and the sex ratio, in terms of numbers, is usually one to one.

In its long journey down the generations therefore, an average gene will spend approximately half its time sitting in male bodies, and the other half sitting in female bodies. Some gene effects show themselves only in bodies of one sex. These are called sex-limited gene effects. A gene controlling penis-length expresses this effect only in male bodies, but it is carried about in female bodies too and may have some quite different effect on female bodies. There is no reason why a man should not inherit a tendency to develop a long penis from his mother.

In whichever of the two sorts of body it finds itself, we can expect a gene to make the best use of the opportunities offered by that sort of body. These opportunities may well differ according to whether the body is male or female. As a convenient approximation, we can once again assume that each individual body is a selfish machine, trying to do the best for all its genes. The best policy for such a selfish machine will often be one thing if it is male, and quite a different thing if it is female. For brevity, we shall again use the convention of thinking of the individual as though it had a conscious purpose. As before, we shall hold in the back of our mind that this is just a figure of speech. A body is really a machine blindly programmed by its selfish genes.

Consider again the mated pair with which we began the chapter. Both partners, as selfish machines, 'want' sons and daughters in equal numbers. To this extent they agree. Where they disagree is in who is going to bear the brunt of the cost of rearing each one of those children. Each individual wants as many surviving children as possible. The less he or she is obliged to invest in any one of those children, the more children he or she can have. The obvious way to achieve this desirable state of affairs is to induce your sexual partner to invest more than his or her fair share of resources in each child, leaving you free to have other children with other partners. This would be a desirable strategy for either sex, but it is more difficult for the female to achieve. Since she starts by investing more than the male, in the form of her large, food-rich egg, a mother is already at the moment of conception 'committed' to each child more deeply than the father is. She stands to lose more if the child dies than the father does. More to the point, she would have to invest more than the father *in the future* in order to bring a new substitute child up to the same level of development. If she tried the tactic of leaving the father holding the baby, while she went off with another male, the father might, at relatively small cost to himself, retaliate by abandoning the baby too. Therefore, at least in the early stages of child development, if any abandoning is going to be done, it is likely to be the father who abandons the mother rather than the other way around. Similarly, females can be expected to invest more in children than males, not only at the outset, but throughout development. So, in mammals for example, it is the female who incubates the foetus in her own body, the female who makes the milk to suckle it when it is born, the female who bears the brunt of the load of bringing it up and protecting it. The female sex is exploited, and the fundamental evolutionary basis for the exploitation is the fact that eggs are larger than sperms.

Of course in many species the father does work hard and faithfully at looking after the young. But even so, we must expect that there will normally be some evolutionary pressure on males to invest a little bit less in each child, and to try to have more children by different wives. By this I simply mean that there will be a tendency for genes which say 'Body, if you are male leave your mate a little bit earlier than my rival allele would have you do, and look for another female', to be successful in the gene pool. The extent to which this evolutionary pressure actually prevails in practice varies greatly from species to species. In many, for example in the birds of paradise, the female receives no help at all from any male, and she rears her children on her own. Other species such as kittiwakes form monogamous pair-bonds of exemplary fidelity, and both partners cooperate in the work of bringing up children. Here we must suppose that some evolutionary counter-pressure have been at work: there must be a penalty attached to the selfish mate-exploitation strategy as well as a benefit, and in kittiwakes the penalty outweighs the benefit. It will in any case only pay a father to desert his wife and child if the wife has a reasonable chance of rearing the child on her own.

Trivers has considered the possible courses of action open to a mother who has been deserted by her mate. Best of all for her would be to try to deceive another male into adopting her child, 'thinking' it is his own. This might not be too difficult if it is still a foetus, not yet born. Of course, while the child bears half her genes, it bears no genes at all from the gullible step-father.

Natural selection would severely penalize such gullibility in males and indeed would favour males who took active steps to kill any potential step-children as soon as they mated with a new wife. This is very probably the explanation of the so-called Bruce effect: male mice secrete a chemical which when smelt by a pregnant female can cause her to abort. She only aborts if the smell is different from that of her former mate. In this way a male mouse destroys his potential step-children, and renders his new wife receptive to his own sexual advances. Ardrey, incidentally, sees the Bruce effect as a population control mechanism! A similar example is that of male lions, who, when newly arrived in a pride, usually murder all existing cubs, presumably because these are not their own children.

A male can achieve the same result without necessarily killing step-children. He can enforce a period of prolonged courtship before he copulates with a female, driving away all other males who approach her, and preventing her from escaping. In this way he can wait and see whether she is harbouring any little step-children in her womb, and desert her if so. We shall see below a reason why a female might want a long 'engagement' period before copulation. Here we have a reason why a male might want one too. Provided he can isolate her from all contact with other males, it helps to avoid being the unwitting benefactor of another male's children.

Assuming then that a deserted female cannot fool a new male into adopting her child, what else can she do? Much may depend on how old the child is. If it is only just conceived, it is true that she has invested the whole of one egg in it and perhaps more, but it may still pay her to abort it and find a new mate as quickly as possible. In these circumstances it would be to the mutual advantage both of her and of the potential new husband that she should abort—since we are assuming she has no hope of fooling him into adopting the child. This could explain why the Bruce effect works from the female's point of view.

Another option open to a deserted female is to stick it out, and try and rear the child on her own. This will especially pay her if the child is already quite old. The older he is the more has already been invested in him, and the less it will take out of her to finish the job of rearing him. Even if he is still quite young, it might yet pay her to try to salvage something from her initial investment, even if she has to work twice as hard to feed the child, now that the male has gone. It is no comfort to her that the child contains half the male's genes too, and that she could spite him by abandoning it. There is no point in spite for its own sake. The child carries half her genes, and the dilemma is now hers alone.

Paradoxically, a reasonable policy for a female who is in danger of being deserted might be to walk out on the male *before* he walks out on her. This could pay her, even if she has already invested more in the child than the male has. The unpleasant truth is that in some circumstances an advantage accrues to the partner who deserts *first*, whether it is the father or the mother. As Trivers puts it, the partner who is left behind is placed in a cruel bind. It is a rather horrible but very subtle argument. A parent may be expected to desert, the moment it is possible for him or her to say the following: 'This child is now far enough developed that either of us *could* finish off rearing it on our own. Therefore it would pay me to desert now, provided I could be sure my partner would not desert as well. If I did desert now, my partner would do whatever is best for her/his genes. He/she would be forced into making a more drastic decision than I am making now, because I would have already left. My partner would "know" that if he/she left as well, the child would surely die. Therefore, assuming that my partner will take the decision which is best for his/her own selfish genes, I conclude that my own best course of action is to desert first. This is especially so, since my partner may be "thinking" along exactly the same lines, and may seize the initiative at any minute by deserting me!' As always, the subjective soliloquy is intended for illustration only. The point is that genes for deserting *first* could be favourably selected simply because genes for deserting *second* would not be.

We have looked at some of the things which a female might do if she has been deserted by her mate. But these all have the air of making the best of a bad job. Is there anything a female can do to reduce the extent to which her mate exploits her in the first place? She has a strong card in her hand. She can refuse to copulate. She is in demand, in a seller's market. This is because she brings the dowry of a large, nutritious egg. A male who successfully copulates gains a valuable food reserve for his offspring. The female is potentially in a position to drive a hard bargain before she copulates. Once she has copulated she has played her ace— her egg has been committed to the male. It is all very well to talk about driving hard bargains, but we know very well it is not really like that. Is there any realistic way in which something equivalent to driving a hard bargain could evolve by natural selection? I shall consider two main possibilities, called the domestic-bliss strategy, and the he-man strategy.

The simplest version of the domestic-bliss strategy is this. The female looks the males over, and tries to spot signs of fidelity and domesticity in advance. There is bound to be variation in the population of males in their predisposition to be faithful husbands. If females could recognize such qualities in advance, they could benefit themselves by choosing males possessing them. One way for a female to do this is to play hard to get for a long time, to be coy. Any male who is not patient enough to wait until the female eventually consents to copulate is not likely to be a good bet as a faithful husband. By insisting on a long engagement period, a female weeds out casual suitors, and only finally copulates with a male who has proved his qualities of fidelity and perseverance in advance. Feminine coyness is in fact very common among animals, and so are prolonged courtship or engagement periods. As we have already seen, a long engagement can also benefit a male where there is a danger of his being duped into caring for another male's child.

Courtship rituals often include considerable pre-copulation investment by the male. The female may refuse to copulate until the male has built her a nest. Or the male may have to feed her quite substantial amounts of food. This, of course, is very good from the female's point of view, but it also suggests another possible version of the domestic-bliss strategy. Could females force males to invest so heavily in their offspring *before* they allow copulation that it would no longer pay the males to desert *after* copulation? The idea is appealing. A male who waits for a coy female eventually to copulate with him is paying a cost: he is forgoing the chance to copulate with other females, and he is spending a lot of time and energy in courting her. By the time he is finally allowed to copulate with a particular female, he will inevitably be heavily 'committed' to her. There will be little temptation for him to desert her, if he knows that any future female he approaches will also procrastinate in the same manner before she will get down to business.

As I showed in a paper, there is a mistake in Trivers's reasoning here. He thought that prior investment in itself committed an individual to future investment. This is fallacious economics. A business man should never say 'I have already invested so much in the Concorde airliner (for instance) that I cannot afford to scrap it now.' He should always ask instead whether it would pay him *in the future*, to cut his losses, and abandon the project now, even though he has already invested heavily in it. Similarly, it is no use a female forcing a male to invest heavily in her in the hope that this, on its own, will deter the male from subsequently deserting. This version of the domestic-bliss strategy depends upon one further crucial assumption. This is that a majority of the females can be relied upon to play the same game. If there are loose females in the population, prepared to welcome males who have deserted their wives, then it could pay a male to desert his wife, no matter how much he has already invested in her children.

Much therefore depends on how the majority of females behave. If we were allowed to think in terms of a conspiracy of females there would be no problem. But a conspiracy of females can no more evolve than the conspiracy of doves which we considered in Chapter 5. Instead, we must look for evolutionarily stable strategies. Let us take Maynard Smith's method of analysing aggressive contests, and apply it to sex. It will be a little bit more complicated than the case of the hawks and doves, because we shall have two female strategies and two male strategies.

As in Maynard Smith's analysis, the word 'strategy' refers to a blind unconscious behaviour program. The two female strategies will be called *coy* and *fast*, and the two male strategies will be called *faithful* and *philanderer*. The behavioural rules of the four types are as follows. Coy females will not copulate with a male until he has gone through a long and expensive courtship period lasting several weeks. Fast females will copulate immediately with anybody. Faithful males are prepared to go on courting for a long time, and after copulation they stay with the female and help her to rear the young. Philanderer males lose patience quickly if a female will not copulate with them straight away: they go off and look for another female; after copulation too they do not stay and act as good fathers, but go off in search of fresh females. As in the case of the hawks and doves, these are not the only possible strategies, but it is illuminating to study their fates nevertheless.

Like Maynard Smith, we shall use some arbitrary hypothetical values for the various costs and benefits. To be more general it can be done with algebraic symbols, but numbers are easier to understand. Suppose that the genetic pay-off gained by each parent when a child is reared successfully is +15 units. The cost of rearing one child, the cost of all its food, all the time spent looking after it, and all the risks taken on its behalf, is −20 units. The cost is expressed as negative, because it is 'paid out' by the parents. Also negative is the cost of wasting time in prolonged courtship. Let this cost be −3 units.

Imagine we have a population in which all the females are coy, and all the males are faithful. It is an ideal monogamous society. In each couple, the male and the female both get the same average pay-off. They get +15 for each child reared; they share the cost of rearing it (−20) equally between the two of them, an average of −10 each. They both pay the −3 point penalty for wasting time in prolonged courtship. The average pay-off for each is therefore +15 − 10 − 3 = +2.

Now suppose a single fast female enters the population. She does very well. She does not pay the cost of delay, because she does not indulge in prolonged courtship. Since all the males in the population are faithful, she can reckon on finding a good father for her children whoever she mates with. Her average pay-off per child is +15 − 10 = +5. She is 3 units better off than her coy rivals. Therefore fast genes will start to spread.

If the success of fast females is so great that they come to predominate in the population, things will start to change in the male camp too. So far, faithful males have had a monopoly. But now if a philanderer male arises in the population, he starts to do better than his faithful rivals. In a population where all the females are fast, the pickings for a philanderer male are rich indeed. He gets the +15 points if a child is successfully reared, and he pays neither of the two costs. What this lack of cost mainly means to him is that he is free to go off and mate with new females. Each of his unfortunate wives struggles on alone with the child, paying the entire −20 point cost, although she does not pay anything for wasting time in courting. The net pay-off for a fast female when she encounters a philanderer male is +15 − 20 = −5; the pay-off to the philanderer himself is +15. In a population in which all the females are fast, philanderer genes will spread like wildfire.

If the philanderers increase so successfully that they come to dominate the male part of the population, the fast females will be in dire straits. Any coy female would have a strong advantage. If a coy female encounters a philanderer male, no business results. She insists on prolonged courtship; he refuses and goes off in search of another female. Neither partner pays the cost of wasting time. Neither gains anything either, since no child is produced. This gives a net pay-off of zero for a coy female in a population where all the males are philanderers. Zero may not seem much, but it is better than the −5 which is the average score for a fast female. Even if a fast female decided to leave her young after being deserted by a philanderer, she would still have paid the considerable cost of an egg. So, coy genes start to spread through the population again.

To complete the hypothetical cycle, when coy females increase in numbers so much that they predominate, the philanderer males, who had such an easy time with the fast females, start to feel the pinch. Female after female insists on a long and ardous courtship. The philanderers flit from female to female, and always the story is the same. The net pay-off for a philanderer male when all the females are coy is zero. Now if a single faithful male should turn up, he is the only one with whom the coy females will mate. His net pay-off is +2, better than that of the philanderers. So, faithful genes start to increase, and we come full circle.

As in the case of the aggression analysis, I have told the story as though it was an endless oscillation. But, as in that case, it can be shown that really there would be no oscillation. The system would converge to a stable state. If you do the sums, it turns out that a population in which $\frac{5}{6}$ of the females are coy, and $\frac{5}{8}$ of the males are faithful, is evolutionarily stable. This is, of course, just for the particular arbitrary numbers which we started out with, but it is easy to work out what the stable ratios would be for any other arbitrary assumptions.

As in Maynard Smith's analyses, we do not have to think of there being two different sorts of male and two different sorts of female. The ESS could equally well be achieved if each male spends $\frac{5}{8}$ of his time being faithful and the rest of his time philandering; and each female spends $\frac{5}{6}$ of her time being coy and $\frac{1}{6}$ of her time being fast. Whichever way we think of the ESS, what it means is this. Any tendency for members of either sex to deviate from their appropriate stable ratio will be penalized by a consequent change in the ratio of strategies of the other sex, which is, in turn, to the disadvantage of the original deviant. Therefore the ESS will be preserved.

We can conclude that it is certainly possible for a population consisting largely of coy females and philanderer males to evolve. In these circumstances the domestic-bliss strategy for females really does seem to work. We do not have to think in terms of a conspiracy of coy females. Coyness can actually pay a female's selfish genes.

There are various ways in which females can put this type of strategy into practice. I have already suggested that a female might refuse to copulate with a male who has not already built her a nest, or at least helped her to build a nest. It is indeed the case that in many monogamous birds copulation does not take place until after the nest is built. The effect of this is that at the moment of conception the male has invested a good deal more in the child than just his cheap sperms.

Demanding that a prospective mate should build a nest is one effective way for a female to trap him. It might be thought that almost anything that costs the male a great deal would do in theory, even if that cost is not directly paid in the form of benefit to the unborn children. If all females of a population forced males to do some difficult and costly deed, like slaying a dragon or climbing a mountain, before they would consent to copulate with them, they could in theory be reducing the temptation for the males to desert after copulation. Any male tempted to desert his mate and try to spread more of his genes by another female, would be put off by the thought that he would have to kill another dragon. In practice, however, it is unlikely that females would impose such arbitrary tasks as dragon-killing, or Holy-Grail-seeking on their suitors. The reason is that a rival female who imposed a task no less arduous, but more useful to her and her children, would have an advantage over more romantically minded females who demanded a pointless labour of love. Building a nest may be less romantic than slaying a dragon or swimming the Hellespont, but it is much more useful.

Also useful to the female is the practice I have already mentioned of courtship feeding by the male. In birds this has usually been regarded as a kind of regression to juvenile behaviour on the part of the female. She begs from the male, using the same gestures as a young bird would use. It has been supposed that this is automatically attractive to the male, in the same way as a man finds a lisp or pouting lips attractive in an adult woman. The female bird at this time needs all the extra food she can get, for she is building up her reserves for the effort of manufacturing her enormous eggs. Courtship feeding by the male probably represents direct investment by him in the eggs themselves. It therefore has the effect of reducing the disparity between the two parents in their initial investment in the young.

Several insects and spiders also demonstrate the phenomenon of courtship feeding. Here an alternative interpretation has sometimes been only too obvious. Since, as in the case of the praying mantis, the male may be in danger of being eaten by the larger female, anything which he can do to reduce her appetite may be to his advantage. There is a macabre sense in which the unfortunate male mantis can be said to invest in his children. He is used as food to help make the eggs which will then be fertilized, posthumously, by his own stored sperms.

A female, playing the domestic-bliss strategy, who simply looks the males over and tries to *recognize* qualities of fidelity in advance, lays herself open to deception. Any male who can pass himself off as a good loyal domestic type, but who in reality is concealing a strong tendency towards desertion and unfaithfulness, could have a great advantage. As long as his deserted former wives have any chance of bringing up some of the children, the philanderer stands to pass on more genes than a rival male who is an honest husband and father. Genes for effective deception by males will tend to be favoured in the gene pool.

Conversely, natural selection will tend to favour females who become good at seeing through such deception. One way they can do this is to play especially hard to get when they are courted by a new male, but in successive breeding seasons to be increasingly ready to accept quickly the advances of last year's mate. This will automatically penalize young males embarking on their first breeding season, whether they are deceivers or not. The brood of naïve first year females would tend to contain a relatively high proportion of genes from unfaithful fathers, but faithful fathers have the advantage in the second and subsequent years of a mother's life, for they do not have to go through the same prolonged energy-wasting and time-consuming courtship rituals. If the majority of individuals in a population are the children of experienced rather than naïve mothers—a reasonable assumption in any long-lived species—genes for honest, good fatherhood will come to prevail in the gene pool.

For simplicity, I have talked as though a male were either purely honest or thoroughly deceitful. In reality it is more probable that all males, indeed all individuals, are a little bit deceitful, in that they are programmed to take advantage of opportunities to exploit their mates. Natural selection, by sharpening up the ability of each partner to detect dishonesty in the other, has kept large-scale deceit down to a fairly low level. Males have more to gain from dishonesty than females, and we must expect that, even in those species where males show considerable parental altruism, they will usually tend to do a bit less work than the females, and to be a bit more ready to abscond. In birds and mammals this is certainly normally the case.

There are species, however, in which the male actually does more work in caring for the children than the female does. Among birds and mammals these cases of paternal devotion are exceptionally rare, but they are common among fish. Why? This is a challenge for the selfish gene theory which has puzzled me for a long time. An ingenious solution was recently suggested to me in a tutorial by Miss T. R. Carlisle. She makes use of Trivers's 'cruel bind' idea, referred to above, as follows.

Many fish do not copulate, but instead simply spew out their
sex cells into the water. Fertilization takes place in the open
water, not inside the body of one of the partners. This is probably
how sexual reproduction first began. Land animals like birds,
mammals, and reptiles, on the other hand, cannot afford this kind
of external fertilization, because their sex cells are too vulnerable
to drying-up. The gametes of one sex—the male, since sperms
are mobile—are introduced into the wet interior of a member of
the other sex—the female. So much is just fact. Now comes the
idea. After copulation, the land-dwelling female is left in physical
possession of the embryo. It is inside her body. Even if she lays
the fertilized egg almost immediately, the male still has time to
vanish, thereby forcing the female into Trivers's 'cruel bind'. The
male is inevitably provided with an opportunity to take the prior
decision to desert, closing the female's options, and forcing her to
decide whether to leave the young to certain death, or whether to
stay with it and rear it. Therefore, maternal care is more common
among land animals than paternal care.

But for fish and other water-dwelling animals things are very
different. If the male does not physically introduce his sperms
into the female's body there is no necessary sense in which the
female is left 'holding the baby'. Either partner might make a
quick getaway and leave the other one in possession of the newly
fertilized eggs. But there is even a possible reason why it might
often be the male who is most vulnerable to being deserted. It
seems probable that an evolutionary battle will develop over who
sheds their sex cells first. The partner who does so has the advantage that he or she can then leave the other one in possession of
the new embryos. On the other hand, the partner who spawns
first runs the risk that his prospective partner may subsequently
fail to follow suit. Now the male is more vulnerable here, if only
because sperms are lighter and more likely to diffuse than eggs. If
a female spawns too early, i.e. before the male is ready, it will
not greatly matter because the eggs, being relatively large and
heavy, are likely to stay together as a coherent clutch for some
time. Therefore a female fish can afford to take the 'risk' of
spawning early. The male dare not take this risk, since if he
spawns too early his sperms will have diffused away before the
female is ready, and she will then not spawn herself, because it

will not be worth her while to do so. Because of the diffusion problem, the male must wait until the female spawns, and then he must shed his sperms over the eggs. But she has had a precious few seconds in which to disappear, leaving the male in possession, and forcing him on to the horns of Trivers's dilemma. So this theory neatly explains why paternal care is common in water but rare on dry land.

Leaving fish, I now turn to the other main female strategy, the he-man strategy. In species where this policy is adopted the females, in effect, resign themselves to getting no help from the father of their children, and go all-out for good genes instead. Once again they use their weapon of withholding copulation. They refuse to mate with just any male, but exercise the utmost care and discrimination before they will allow a male to copulate with them. Some males undoubtedly do contain a larger number of good genes than other males, genes which would benefit the survival prospects of both sons and daughters. If a female can somehow detect good genes in males, using externally visible clues, she can benefit her own genes by allying them with good paternal genes. To use our analogy of the rowing crews, a female can minimize the chance that her genes will be dragged down through getting into bad company. She can try to hand-pick good crew-mates for her own genes.

The chances are that most of the females will agree with each other on which are the best males, since they all have the same information to go on. Therefore these few lucky males will do most of the copulating. This they are quite capable of doing, since all they must give to each female is some cheap sperms. This is presumably what has happened in elephant seals and in birds of paradise. The females are allowing just a few males to get away with the ideal selfish-exploitation strategy which all males aspire to, but they are making sure that only the best males are allowed this luxury.

From the point of view of a female trying to pick good genes with which to ally her own, what is she looking for? One thing she wants is evidence of ability to survive. Obviously any potential mate who is courting her has proved his ability to survive at least into adulthood, but he has not necessarily proved that he can survive much longer. Quite a good policy for a female might be to go for old men. Whatever their shortcomings, they have at least proved they can survive, and she is likely to be allying her genes with genes for longevity. However, there is no point in ensuring that her children live long lives if they do not also give her lots of grandchildren. Longevity is not prima facie evidence of virility. Indeed a long-lived male may have survived precisely *because* he does not take risks in order to reproduce. A female who selects an old male is not necessarily going to have more descendants than a rival female who chooses a young one who shows some other evidence of good genes.

What other evidence? There are many possibilities. Perhaps strong muscles as evidence of ability to catch food, perhaps long legs as evidence of ability to run away from predators. A female might benefit her genes by allying them with such traits, since they might be useful qualities in both her sons and her daughters. To begin with, then, we have to imagine females choosing males on the basis of perfectly genuine labels or indicators which tend to be evidence of good underlying genes. But now here is a very interesting point, realized by Darwin, and clearly enunciated by Fisher. In a society where males compete with each other to be chosen as he-men by females, one of the best things a mother can do for her genes is to make a son who will turn out in his turn to be an attractive he-man. If she can ensure that her son is one of the fortunate few males who wins most of the copulations in the society when he grows up, she will have an enormous number of grandchildren. The result of this is that one of the most desirable qualities a male can have in the eyes of a female is, quite simply, sexual attractiveness itself. A female who mates with a super-attractive he-man is more likely to have sons who are attractive to females of the next generation, and who will make lots of grandchildren for her. Originally, then, females may be thought of as selecting males on the basis of obviously useful qualities like big muscles, but once such qualities became widely accepted as attractive among the females of a species, natural selection would continue to favour them simply because they were attractive.

Extravagances such as the tails of male birds of paradise may therefore have evolved by a kind of unstable, run-away process. In the early days, a slightly longer tail than usual may have been selected by females as a desirable quality in males, perhaps because it betokened a fit and healthy constitution. A short tail on a male might have been an indicator of some vitamin deficiency—evidence of poor food-getting ability. Or perhaps short-tailed males were not very good at running away from predators, and so had had their tails bitten off. Notice that we don't have to assume that the short tail was in itself genetically inherited, only that it served as an indicator of some genetic inferiority. Anyway, for whatever reason, let us suppose that females in the ancestral bird of paradise species preferentially went for males with longer than average tails. Provided there was *some* genetic contribution to the natural variation in male tail-length, this would in time cause the average tail-length of males in the population to increase. Females followed a simple rule: look all the males over, and go for the one with the longest tail. Any female who departed from this rule was penalized, *even if* tails had already become so long that they actually encumbered males possessing them. This was because any female who did not produce long-tailed sons had little chance of one of her sons being regarded as attractive. Like a fashion in womens' clothes, or in American car design, the trend toward longer tails took off and gathered its own momentum. It was stopped only when tails became so grotesquely long that their manifest disadvantages started to outweigh the advantage of sexual attractiveness.

This is a hard idea to swallow, and it has attracted its sceptics ever since Darwin first proposed it, under the name of sexual selection. One person who does not believe it is A. Zahavi, whose 'Fox, fox' theory we have already met. He puts forward his own maddeningly contrary 'handicap principle' as a rival explanation. He points out that the very fact that females are trying to select for good genes among males opens the door to deception by the males. Strong muscles may be a genuinely good quality for a female to select, but then what is to stop males from growing dummy muscles with no more real substance than human padded shoulders? If it costs a male less to grow false muscles than real ones, sexual selection should favour genes for producing false muscles. It will not be long, however, before counter-selection leads to the evolution of females capable of seeing through the deception. Zahavi's basic premise is that false sexual advertisement will eventually be seen through by females. He therefore concludes that really successful males will be those who do not advertise falsely, those who palpably demonstrate that they are not deceiving. If it is strong muscles we are talking about, then males who merely assume the visual *appearance* of strong muscles will soon be detected by the females. But a male who demonstrates, by the equivalent of lifting weights or ostentatiously doing press-ups, that he really has strong muscles, will succeed in convincing the females. In other words Zahavi believes that a he-man must not only *seem* to be a good quality male: he must really *be* a good quality male, otherwise he will not be accepted as such by sceptical females. Displays will therefore evolve which only a genuine he-man is capable of doing.

So far so good. Now comes the part of Zahavi's theory which really sticks in the throat. He suggests that the tails of birds of paradise and peacocks, the huge antlers of deer, and the other sexually-selected features which have always seemed paradoxical because they appear to be handicaps to their possessors, evolve precisely *because* they are handicaps. A male bird with a long and cumbersome tail is showing off to females that he is such a strong he-man that he can survive *in spite of* his tail. Think of a woman watching two men run a race. If both arrive at the finishing post at the same time, but one has deliberately encumbered himself with a sack of coal on his back, the women will naturally draw the conclusion that the man with the burden is really the faster runner.

I do not believe this theory, although I am not quite so confident in my scepticism as I was when I first heard it. I pointed out then that the logical conclusion to it should be the evolution of males with only one leg and only one eye. Zahavi, who comes from Israel, instantly retorted: 'Some of our best generals have only one eye!' Nevertheless, the problem remains that the handicap theory seems to contain a basic contradiction. If the handicap is a genuine one—and it is of the essence of the theory that it has to be a genuine one—then the handicap itself will penalize the offspring just as surely as it may attract females. It is, in any case, important that the handicap must not be passed on to daughters.

If we rephrase the handicap theory in terms of genes, we have something like this. A gene which makes males develop a handicap, such as a long tail, becomes more numerous in the gene pool because females choose males who have handicaps. Females choose males who have handicaps, because genes which make females so choose also become frequent in the gene pool. This is because females with a taste for handicapped males will automatically tend to be selecting males with good genes in other respects, since those males have survived to adulthood in spite of the handicap. These good 'other' genes will benefit the bodies of the children, which therefore survive to propagate the genes for the handicap itself, and also the genes for choosing handicapped males. Provided the genes for the handicap itself exert their effect only in sons, just as the genes for a sexual preference for the handicap affect only daughters, the theory just might be made to work. So long as it is formulated only in words, we cannot be sure whether it will work or not. We get a better idea of how feasible such a theory is when it is rephrased in terms of a mathematical model. So far mathematical geneticists who have tried to make the handicap principle into a workable model have failed. This may be because it is not a workable principle, or it may be because they are not clever enough. One of them is Maynard Smith, and my hunch favours the former possibility.

If a male can demonstrate his superiority over other males in a way that does not involve deliberately handicapping himself, nobody would doubt that he could increase his genetic success in that way. Thus elephant seals win and hold on to their harems, not by being aesthetically attractive to females, but by the simple expedient of beating up any male who tries to move in on the harem. Harem holders tend to win these fights against would-be usurpers, if only for the obvious reason that that is why they are harem-holders. Usurpers do not often win fights, because if they were capable of winning they would have done so before! Any female who mates only with a harem holder is therefore allying her genes with a male who is strong enough to beat off successive challenges from the large surplus of desperate bachelor males. With luck her sons will inherit their father's ability to hold a harem. In practice a female elephant seal does not have much option, because the harem-owner beats *her* up if she tries to stray. The principle remains, however, that females who choose to mate with males who win fights may benefit their genes by so doing. As we have seen, there are examples of females preferring to mate with males who hold territories and with males who have high status in the dominance hierarchy.

To sum up this chapter so far, the various different kinds of breeding system that we find among animals—monogamy, promiscuity, harems, and so on—can be understood in terms of conflicting interests between males and females. Individuals of either sex 'want' to maximize their total reproductive output during their lives. Because of a fundamental difference between the size and numbers of sperms and eggs, males are in general likely to be biased towards promiscuity and lack of paternal care. Females have two main available counter-ploys, which I have called the he-man and the domestic-bliss strategies. The ecological circumstances of a species will determine whether the females are biased towards one or the other of these counter-ploys, and will also determine how the males respond. In practice all intermediates between he-man and domestic-bliss are found and, as we have seen, there are cases in which the father does even more child-care than the mother. This book is not concerned with the details of particular animal species, so I will not discuss what might predispose a species towards one form of breeding system rather than another. Instead I will consider the differences which are commonly observed between males and females in general, and show how these may be interpreted. I shall therefore not be emphasizing those species in which the differences between the sexes are slight, these being in general the ones whose females have favoured the domestic-bliss strategy.

Firstly, it tends to be the males who go in for sexually attractive, gaudy colours, and the females who tend to be more drab. Individuals of both sexes want to avoid being eaten by predators, and there will be some evolutionary pressure on both sexes to be drably coloured. Bright colours attract predators no less than they attract sexual partners. In gene terms, this means that genes for bright colours are more likely to meet their end in the stomachs of predators than are genes for drab colours. On the other hand, genes for drab colours may be less likely than genes for bright colours to find themselves in the next generation, because drab individuals have difficulty in attracting a mate. There are therefore two conflicting selection pressures: predators tending to remove bright-colour genes from the gene pool, and sexual partners tending to remove genes for drabness. As in so many other cases, efficient survival machines can be regarded as a compromise between conflicting selection pressures. What interests us at the moment is that the optimal compromise for a male seems to be different from the optimal compromise for a female. This is of course fully compatible with our view of males as high-risk, high-reward gamblers. Because a male produces many millions of sperms to every egg produced by a female, sperms heavily out-number eggs in the population. Any given egg is therefore much more likely to enter into sexual fusion than any given sperm is. Eggs are a relatively valuable resource, and therefore a female does not need to be so sexually attractive as a male does in order to ensure that her eggs are fertilized. A male is perfectly capable of siring all the children born to a large population of females. Even if a male has a short life because his gaudy tail attracts predators, or gets tangled in the bushes, he may have fathered a very large number of children before he dies. An unattractive or drab male may live even as long as a female, but he has few children, and his genes are not passed on. What shall it profit a male if he shall gain the whole world, and lose his immortal genes?

Another common sexual difference is that females are more fussy than males about whom they mate with. One of the reasons for fussiness by an individual of either sex is the need to avoid mating with a member of another species. Such hybridizations are a bad thing for a variety of reasons. Sometimes, as in the case of a man copulating with a sheep, the copulation does not lead to an embryo being formed, so not much is lost. When more closely related species like horses and donkeys cross-breed, however, the cost, at least to the female partner, can be considerable. An embryo mule is likely to be formed, and it then clutters up her womb for eleven months. It takes a large quantity of her total parental investment, not only in the form of food absorbed through the placenta, and then later in the form of milk, but above all in time which could have been spent in rearing other children. Then when the mule reaches adulthood it turns out to be sterile. This is presumably because, although horse chromosomes and donkey chromosomes are sufficiently similar to cooperate in the building of a good strong mule body, they are not similar enough to work together properly in meiosis. Whatever the exact reason, the very considerable investment by the mother in the rearing of a mule is totally wasted from the point of view of her genes. Female donkeys should be very, very careful that the individual they copulate with is another donkey, and not a horse. In gene terms, any donkey gene which says 'Body, if you are female, copulate with any old male, whether he is a donkey or a horse', is a gene which may next find itself in the dead-end body of a mule, and the mother's parental investment in that baby mule detracts heavily from her capacity to rear fertile donkeys. A male, on the other hand, has less to lose if he mates with a member of the wrong species, and, although he may have nothing to gain either, we should expect males to be less fussy in their choice of sexual partners. Where this has been looked at, it has been found to be true.

Even within a species, there may be reasons for fussiness. Incestuous mating, like hybridization, is likely to have damaging genetic consequences, in this case because lethal and semi-lethal recessive genes are brought out into the open. Once again, females have more to lose than males, since their investment in any particular child tends to be greater. Where incest taboos exist, we should expect females to be more rigid in their adherence to the taboos than males. If we assume that the older partner in an incestuous relationship is relatively likely to be the active initiator, we should expect that incestuous unions in which the male is older than the female should be more common than unions in which the female is older. For instance father/daughter incest should be commoner than mother/son. Brother/sister incest should be intermediate in commonness.

In general, males should tend to be more promiscuous than females. Since a female produces a limited number of eggs at a relatively slow rate, she has little to gain from having a large number of copulations with different males. A male on the other hand, who can produce millions of sperms every day, has everything to gain from as many promiscuous matings as he can snatch. Excess copulations may not actually cost a female much, other than a little lost time and energy, but they do not do her positive good. A male on the other hand can never get enough copulations with as many different females as possible: the word excess has no meaning for a male.

I have not explicitly talked about man but inevitably, when we think about evolutionary arguments such as those in this chapter, we cannot help reflecting about our own species and our own experience. Notions of females withholding copulation until a male shows some evidence of long-term fidelity may strike a familiar chord. This might suggest that human females play the domestic-bliss rather than the he-man strategy. Most human societies are indeed monogamous. In our own society, parental investment by both parents is large and not obviously unbalanced. Mothers certainly do more direct work for children than fathers do, but fathers often work hard in a more indirect sense to provide the material resources that are poured into the children. On the other hand, some human societies are promiscuous, and some are harem-based. What this astonishing variety suggests is that man's way of life is largely determined by culture rather than by genes. However, it is still possible that human males in general have a tendency towards promiscuity, and females a tendency towards monogamy, as we would predict on evolutionary grounds. Which of these two tendencies wins in particular societies depends on details of cultural circumstance, just as in different animal species it depends on ecological details.

One feature of our own society which seems decidedly anomalous is the matter of sexual advertisement. As we have seen, it is strongly to be expected on evolutionary grounds that, where the sexes differ, it should be the males who advertise and the females who are drab. Modern western man is undoubtedly exceptional in this respect. It is of course true that some men dress flamboyantly and some women dress drably but, on average, there can be no doubt that in our society the equivalent of the peacock's tail is exhibited by the female, not by the male. Women paint their faces and glue on false eyelashes. Apart from actors and homosexuals, men do not. Women seem to be interested in their own personal appearance and they are encouraged in this by their magazines and journals. Men's magazines are less preoccupied with male sexual attractiveness, and a man who is unusually interested in his own dress and appearance is apt to arouse suspicion, both among men and among women. When a woman is described in conversation, it is quite likely that her sexual attractiveness, or lack of it, will be prominently mentioned. This is true, whether the speaker is a man or a woman. When a man is described, the adjectives used are much more likely to have nothing to do with sex.

Faced with these facts, a biologist would be forced to suspect that he was looking at a society in which females compete for males, rather than vice versa. In the case of birds of paradise, we decided that females are drab because they do not need to compete for males. Males are bright and ostentatious because females are in demand and can afford to be choosy. The reason female birds of paradise are in demand is that eggs are a more scarce resource than sperms. What has happened in modern western man? Has the male really become the sought-after sex, the one that is in demand, the sex that can afford to be choosy? If so, why?

10. You scratch my back, I'll ride on yours

WE have considered parental, sexual, and aggressive interactions between survival machines belonging to the same species. There are striking aspects of animal interactions which do not seem to be obviously covered by any of these headings. One of these is the propensity which so many animals have for living in groups. Birds flocks, insects swarm, fish and whales school, plains-dwelling mammals herd together or hunt in packs. These aggregations usually consist of members of a single species only, but there are exceptions. Zebras often herd together with gnus, and mixed-species flocks of birds are sometimes seen.

The suggested benefits which a selfish individual can wrest from living in a group constitute rather a miscellaneous list. I am not going to trot out the catalogue, but will mention just a few suggestions. In the course of this I will return to the remaining examples of apparently altruistic behaviour which I gave in Chapter 1, and which I promised to explain. This will lead into a consideration of the social insects, without which no account of animal altruism would be complete. Finally in this rather miscellaneous chapter, I shall mention the important idea of reciprocal altruism, the principle of 'You scratch my back, I'll scratch yours'.

If animals live together in groups their genes must get more benefit out of the association than they put in. A pack of hyenas can catch prey so much larger than a lone hyena can bring down that it pays each selfish individual to hunt in a pack, even though this involves sharing food. It is probably for similar reasons that some spiders cooperate in building a huge communal web. Emperor penguins conserve heat by huddling together. Each one gains by presenting a smaller surface area to the elements than he would on his own. A fish who swims obliquely behind another fish may gain a hydrodynamic advantage from the turbulence produced by the fish in front. This could be partly why fish school. A related trick concerned with air turbulence is known to racing cyclists, and it may account for the V-formation of flying birds. There is probably competition to avoid the disadvantageous position at the head of the flock. Possibly the birds take turns as unwilling leader—a form of the delayed reciprocal-altruism to be discussed at the end of the chapter.

Many of the suggested benefits of group living have been concerned with avoiding being eaten by predators. An elegant formulation of such a theory was given by W. D. Hamilton, in a paper called *Geometry for the selfish herd*. Lest this lead to misunderstanding, I must stress that by 'selfish herd' he meant 'herd of selfish individuals'.

Once again we start with a simple 'model' which, though abstract, helps us to understand the real world. Suppose a species of animal is hunted by a predator which always tends to attack the nearest prey individual. From the predator's point of view this is a reasonable strategy, since it tends to cut down energy expenditure. From the prey's point of view it has an interesting consequence. It means that each prey individual will constantly try to avoid being the nearest to a predator. If the prey can detect the predator at a distance, it will simply run away. But if the predator is apt to turn up suddenly without warning, say it lurks concealed in long grass, then each prey individual can still take steps to minimize its chance of being the nearest to a predator. We can picture each prey individual as being surrounded by a 'domain of danger'. This is defined as that area of ground in which any point is nearer to that individual than it is to any other individual. For instance, if the prey individuals march spaced out in a regular geometric formation, the domain of danger round each one (unless he is on the edge) might be roughly hexagonal in shape. If a predator happens to be lurking in the hexagonal domain of danger surrounding individual A, then individual A is likely to be eaten. Individuals on the edge of the herd are especially vulnerable, since their domain of danger is not a relatively small hexagon, but includes a wide area on the open side.

Now clearly a sensible individual will try to keep his domain of danger as small as possible. In particular, he will try to avoid being on the edge of the herd. If he finds himself on the edge he will take immediate steps to move towards the centre. Unfortunately somebody has to be on the edge, but as far as each individual is concerned it is not going to be him! There will be a ceaseless migration in from the edges of an aggregation towards the centre. If the herd was previously loose and straggling, it will soon become tightly bunched as a result of the inward migration. Even if we start our model with no tendency towards aggregation at all, and the prey animals start by being randomly dispersed, the selfish urge of each individual will be to reduce his domain of danger by trying to position himself in a gap between other individuals. This will quickly lead to the formation of aggregations which will become ever more densely bunched.

Obviously, in real life the bunching tendency will be limited by opposing pressures: otherwise all individuals would collapse in a writhing heap! But still, the model is interesting as it shows us that even very simple assumptions can predict aggregation. Other, more elaborate models have been proposed. The fact that they are more realistic does not detract from the value of the simpler Hamilton model in helping us to think about the problem of animal aggregation.

The selfish-herd model in itself has no place for cooperative interactions. There is no altruism here, only selfish exploitation by each individual of every other individual. But in real life there are cases where individuals seem to take active steps to preserve fellow members of the group from predators. Bird alarm calls spring to mind. These certainly function as alarm signals in that they cause individuals who hear them to take immediate evasive action. There is no suggestion that the caller is 'trying to draw the predator's fire' away from his colleagues. He is simply informing them of the predator's existence—warning them. Nevertheless the act of calling seems, at least at first sight, to be altruistic, because it has the *effect* of calling the predator's attention to the caller. We can infer this indirectly from a fact which was noticed by P. R. Marler. The physical characteristics of the calls seem to be ideally shaped to be difficult to locate. If an acoustic engineer were asked to design a sound which a predator would find it hard to approach, he would produce something very like the real alarm calls of many small song-birds. Now in nature this shaping of the calls must have been produced by natural selection, and we know what that means. It means that large numbers of individuals have died because their alarm calls were not quite perfect. Therefore there seems to be danger attached to giving alarm calls. The selfish gene theory has to come up with a convincing advantage of giving alarm calls which is big enough to counteract this danger.

In fact this is not very difficult. Bird alarm calls have been held up so many times as 'awkward' for the Darwinian theory that it has become a kind of sport to dream up explanations for them. As a result, we now have so many good explanations that it is hard to remember what all the fuss was about. Obviously, if there is a chance that the flock contains some close relatives, a gene for giving an alarm call can prosper in the gene pool because it has a good chance of being in the bodies of some of the individuals saved. This is true, even if the caller pays dearly for his altruism by attracting the predator's attention to himself.

If you are not satisfied with this kin-selection idea, there are plenty of other theories to choose from. There are many ways in which the caller could gain selfish benefit from warning his fellows. Trivers reels off five good ideas, but I find the following two of my own rather more convincing.

The first I call the *cave* theory, from the Latin for 'beware', still used (pronounced 'kay-vee') by schoolboys to warn of approaching authority. This theory is suitable for camouflaged birds which crouch frozen in the undergrowth when danger threatens. Suppose a flock of such birds is feeding in a field. A hawk flies past in the distance. He has not yet seen the flock and he is not flying directly towards them, but there is a danger that his keen eyes will spot them at any moment and he will race into the attack. Suppose one member of the flock sees the hawk, but the rest have not yet done so. This one sharp-eyed individual could immediately freeze and crouch in the grass. But this would do him little good, because his companions are still walking around conspicuously and noisily. Any one of them could attract the hawk's attention and then the whole flock is in peril. From a purely selfish point of view the best policy for the individual who spots the hawk first is to hiss a quick warning to his companions, and so shut them up and reduce the chance that they will inadvertently summon the hawk into his own vicinity.

The other theory I want to mention may be called the 'never

break ranks' theory. This one is suitable for species of birds which fly off when a predator approaches, perhaps up into a tree. Once again, imagine that one individual in a flock of feeding birds has spotted a predator. What is he to do? He could simply fly off himself, without warning his colleagues. But now he would be a bird on his own, no longer part of a relatively anonymous flock, but an odd man out. Hawks are actually known to go for odd pigeons out, but even if this were not so there are plenty of theoretical reasons for thinking that breaking ranks might be a suicidal policy. Even if his companions eventually follow him, the individual who first flies up off the ground temporarily increases his domain of danger. Whether Hamilton's particular theory is right or wrong, there must be some important advantage in living in flocks, otherwise the birds would not do it. Whatever that advantage may be, the individual who leaves the flock ahead of the others will, at least in part, forfeit that advantage. If he must not break ranks, then, what is the observant bird to do? Perhaps he should just carry on as if nothing had happened and rely on the protection afforded by his membership of the flock. But this too carries grave risks. He is still out in the open, highly vulnerable. He would be much safer up in a tree. The best policy is indeed to fly up into a tree, *but to make sure everybody else does too.* That way, he will not become an odd man out and he will not forfeit the advantages of being part of a crowd, but he will gain the advantage of flying off into cover. Once again, uttering a warning call is seen to have a purely selfish advantage. E. L. Charnov and J. R. Krebs have proposed a similar theory in which they go so far as to use the word 'manipulation' to describe what the calling bird does to the rest of his flock. We have come a long way from pure, disinterested altruism!

Superficially, these theories may seem incompatible with the statement that the individual who gives the alarm call endangers himself. Really there is no incompatibility. He would endanger himself even more by not calling. Some individuals have died because they gave alarm calls, especially the ones whose calls were easy to locate. Other individuals have died because they did not give alarm calls. The *cave* theory and the 'never break ranks' theory are just two out of many ways of explaining why.

What of the stotting Thomson's gazelle, which I mentioned

in Chapter 1, and whose apparently suicidal altruism moved Ardrey to state categorically that it could be explained only by group selection? Here the selfish gene theory has a more exacting challenge. Alarm calls in birds do work, but they are clearly designed to be as inconspicuous and discreet as possible. Not so the stotting high-jumps. They are ostentatious to the point of downright provocation. The gazelles look as if they are deliberately inviting the predator's attention, almost as if they are teasing the predator. This observation has led to a delightfully daring theory. The theory was originally foreshadowed by N. Smythe but, pushed to its logical conclusion, it bears the unmistakeable signature of A. Zahavi.

Zahavi's theory can be put like this. The crucial bit of lateral thinking is the idea that stotting, far from being a signal to the other gazelles, is really aimed at the predators. It is noticed by the other gazelles and it affects their behaviour, but this is incidental, for it is primarily selected as a signal to the predator. Translated roughly into English it means: 'Look how high I can jump, I am obviously such a fit and healthy gazelle, you can't catch me, you would be much wiser to try and catch my neighbour who is not jumping so high!' In less anthropomorphic terms, genes for jumping high and ostentatiously are unlikely to be eaten by predators because predators tend to choose prey who look easy to catch. In particular, many mammal predators are known to go for the old and the unhealthy. An individual who jumps high is advertising, in an exaggerated way, the fact that he is neither old nor unhealthy. According to this theory, the display is far from altruistic. If anything it is selfish, since its object is to persuade the predator to chase somebody else. In a way there is a competition to see who can jump the highest, the loser being the one chosen by the predator.

The other example that I said I would return to is the case of the kamikaze bees, who sting honey-raiders but commit almost certain suicide in the process. The honey bee is just one example of a highly *social* insect. Others are wasps, ants, and termites or 'white ants'. I want to discuss social insects generally, not just suicidal bees. The exploits of the social insects are legendary, in particular their astonishing feats of cooperation and apparent altruism. Suicidal stinging missions typify their prodigies of self-abnegation. In the 'honey-pot' ants there is a caste of workers with grotesquely swollen, food-packed abdomens, whose sole function in life is to hang motionless from the ceiling like bloated light-bulbs, being used as food stores by the other workers. In the human sense they do not live as individuals at all; their individuality is subjugated, apparently to the welfare of the community. A society of ants, bees, or termites achieves a kind of individuality at a higher level. Food is shared to such an extent that one may speak of a communal stomach. Information is shared so efficiently by chemical signals and by the famous 'dance' of the bees that the community behaves almost as if it were a unit with a nervous system and sense organs of its own. Foreign intruders are recognized and repelled with something of the selectivity of a body's immune reaction system. The rather high temperature inside a beehive is regulated nearly as precisely as that of the human body, even though an individual bee is not a 'warm blooded' animal. Finally and most importantly, the analogy extends to reproduction. The majority of individuals in a social insect colony are sterile workers. The 'germ line'—the line of immortal gene continuity—flows through the bodies of a minority of individuals, the reproductives. These are the analogues of our own reproductive cells in our testes and ovaries. The sterile workers are the analogy of our liver, muscle, and nerve cells.

Kamikaze behaviour and other forms of altruism and cooperation by workers are not astonishing once we accept the fact that they are sterile. The body of a normal animal is manipulated to ensure the survival of its genes both through bearing offspring and through caring for other individuals containing the same genes. Suicide in the interests of caring for other individuals is incompatible with future bearing of one's own offspring. Suicidal self-sacrifice therefore seldom evolves. But a worker bee never bears offspring of its own. All its efforts are directed to preserving its genes by caring for relatives other than its own offspring. The death of a single sterile worker bee is no more serious to its genes than is the shedding of a leaf in autumn to the genes of a tree.

There is a temptation to wax mystical about the social insects, but there is really no need for this. It is worth looking in some detail at how the selfish gene theory deals with them, and in particular at how it explains the evolutionary origin of that extraordinary phenomenon of worker sterility from which so much seems to follow.

A social insect colony is a huge family, usually all descended from the same mother. The workers, who seldom or never reproduce themselves, are often divided into a number of distinct castes, including small workers, large workers, soldiers, and highly specialized castes like the honey-pots. Reproductive females are called queens. Reproductive males are sometimes called drones or kings. In the more advanced societies, the reproductives never work at anything except procreation, but at this one task they are extremely good. They rely on the workers for their food and protection, and the workers are also responsible for looking after the brood. In some ant and termite species the queen has swollen into a gigantic egg factory, scarcely recognizable as an insect at all, hundreds of times the size of a worker and quite incapable of moving. She is constantly tended by workers who groom her, feed her, and transport her ceaseless flow of eggs to the communal nurseries. If such a monstrous queen ever has to move from the royal cell she rides in state on the backs of squadrons of toiling workers.

In Chapter 7, I introduced the distinction between bearing and caring. I said that mixed strategies, combining bearing and caring, would normally evolve. In Chapter 5 we saw that mixed evolutionarily stable strategies could be of two general types. Either each individual in the population could behave in a mixed way: thus individuals usually achieve a judicious mixture of bearing and caring; *or*, the population may be divided into two different types of individual: this was how we first pictured the balance between hawks and doves. Now it is theoretically possible for an evolutionarily stable balance between bearing and caring to be achieved in the latter kind of way: the population could be divided into bearers and carers. But this can only be evolutionarily stable if the carers are close kin to the individuals for whom they care, at least as close as they would be to their own offspring if they had any. Although it is theoretically possible for evolution to proceed in this direction, it seems to be only in the social insects that it has actually happened.

Social insect individuals are divided into two main classes, bearers and carers. The bearers are the reproductive males and females. The carers are the workers—infertile males and females in the termites, infertile females in all other social insects. Both types do their job more efficiently because they do not have to cope with the other. But from whose point of view is it efficient? The question which will be hurled at the Darwinian theory is the familiar cry: 'What's in it for the workers?'

Some people have answered 'Nothing.' They feel that the queen is having it all her own way, manipulating the workers by chemical means to her own selfish ends, making them care for her own teeming brood. This is a version of Alexander's 'parental manipulation' theory which we met in Chapter 8. The opposite idea is that the workers 'farm' the reproductives, manipulating them to increase their productivity in propagating replicas of the workers' genes. To be sure, the survival machines which the queen makes are not offspring to the workers, but they are close relatives nevertheless. It was Hamilton who brilliantly realized that, at least in the ants, bees, and wasps, the workers may actually be more closely related to the brood than the queen herself is! This led him, and later Trivers and Hare, on to one of the most spectacular triumphs of the selfish gene theory. The reasoning goes like this.

Insects of the group known as the Hymenoptera, including ants, bees, and wasps, have a very odd system of sex determination. Termites do not belong to this group and they do not share the same peculiarity. A hymenopteran nest typically has only one mature queen. She made one mating flight when young and stored up the sperms for the rest of her long life—ten years or even longer. She rations the sperms out to her eggs over the years, allowing the eggs to be fertilized as they pass out through her tubes. But not all the eggs are fertilized. The unfertilized ones develop into males. A male therefore has no father, and all the cells of his body contain just a single set of chromosomes (all obtained from his mother) instead of a double set (one from the father and one from the mother) as in ourselves. In terms of the analogy of Chapter 3, a male hymenopteran has only one copy of each 'volume' in each of his cells, instead of the usual two.

A female hymenopteran, on the other hand, is normal in that she does have a father, and she has the usual double set of chromosomes in each of her body cells. Whether a female develops into a worker or a queen depends not on her genes but on how she is brought up. That is to say, each female has a complete set of queen-making genes, and a complete set of worker-making genes (or, rather, sets of genes for making each specialized caste of worker, soldier, etc). Which set of genes is 'turned on' depends on how the female is reared, in particular on the food she receives.

Although there are many complications, this is essentially how things are. We do not know why this extraordinary system of sexual reproduction evolved. No doubt there were good reasons, but for the moment we must just treat it as a curious fact about the Hymenoptera. Whatever the original reason for the oddity, it plays havoc with Chapter 6's neat rules for calculating relatedness. It means that the sperms of a single male, instead of all being different as they are in ourselves, are all exactly the same. A male has only a single set of genes in each of his body cells, not a double set. Every sperm must therefore receive the full set of genes rather than a 50 per cent sample, and all sperms from a given male are therefore identical. Let us now try to calculate the relatedness between a mother and son. If a male is known to possess a gene A, what are the chances that his mother shares it? The answer must be 100 per cent, since the male had no father and obtained all his genes from his mother. But now suppose a queen is known to have the gene B. The chance that her son shares the gene is only 50 per cent, since he contains only half her genes. This sounds like a contradiction, but it is not. A male gets *all* his genes from his mother, but a mother only gives *half* her genes to her son. The solution to the apparent paradox lies in the fact that a male has only half the usual number of genes. There is no point in puzzling over whether the 'true' index of relatedness is $\frac{1}{2}$ or 1. The index is only a man-made measure, and if it leads to difficulties in particular cases, we may have to abandon it and go back to first principles. From the point of view of a gene A in the body of a queen, the chance that the gene is shared by a son is $\frac{1}{2}$, just as it is for a daughter. From a queen's point of view therefore, her offspring, of either sex, are as closely related to her as human children are to their mother.

Things start to get intriguing when we come to sisters. Full sisters not only share the same father: the two sperms which conceived them were identical in every gene. The sisters are therefore equivalent to identical twins as far as their paternal genes are concerned. If one female has a gene A, she must have got it from either her father or her mother. If she got it from her mother then there is a 50 per cent chance that her sister shares it. But if she got it from her father, the chances are 100 per cent that her sister shares it. Therefore the relatedness between hymenopteran full sisters is not $\frac{1}{2}$ as it would be for normal sexual animals, but $\frac{3}{4}$.

It follows that a hymenopteran female is more closely related to her full sisters than she is to her offspring of either sex. As Hamilton realized (though he did not put it in quite the same way) this might well predispose a female to farm her own mother as an efficient sister-making machine. A gene for vicariously making sisters replicates itself more rapidly than a gene for making offspring directly. Hence worker sterility evolved. It is presumably no accident that true sociality, with worker sterility, seems to have evolved no less than eleven times *independently* in the Hymenoptera and only once in the whole of the rest of the animal kingdom, namely in the termites.

However, there is a catch. If the workers are successfully to farm their mother as a sister-producing machine, they must somehow curb her natural tendency to give them an equal number of little brothers as well. From the point of view of a worker, the chance of any one brother containing a particular one of her genes is only $\frac{1}{4}$. Therefore, if the queen were allowed to produce male and female reproductive offspring in equal proportions, the farm would not show a profit as far as the workers are concerned. They would not be maximizing the propagation of their precious genes.

Trivers and Hare realized that the workers must try to bias the sex ratio in favour of females. They took the Fisher calculations on optimal sex ratios (which we looked at in the previous chapter) and re-worked them for the special case of the Hymenoptera. It turned out that the optimal ratio of investment for a mother is, as usual, 1 : 1. But the optimal ratio for a sister is 3 : 1 in favour of sisters rather than brothers. If you are a hymenopteran female, the most efficient way for you to propagate your genes is to refrain from breeding yourself, and to make your mother provide you with reproductive sisters and brothers in the ratio 3 : 1. But if you *must* have offspring of your own, you can benefit your genes best by having reproductive sons and daughters in equal proportions.

As we have seen, the difference between queens and workers is not a genetic one. As far as her genes are concerned, an embryo female might be destined to become either a worker, who 'wants' a 3 : 1 sex ratio, or a queen, who 'wants' a 1 : 1 ratio. So what does this 'wanting' mean? It means that a gene which finds itself in a queen's body can propagate itself best if that body invests equally in reproductive sons and daughters. But the same gene finding itself in a worker's body can propagate itself best by making the mother of that body have more daughters than sons. There is no real paradox here. A gene must take best advantage of the levers of power which happen to be at its disposal. If it finds itself in a position to influence the development of a body which is destined to turn into a queen, its optimal strategy to exploit that control is one thing. If it finds itself in a position to influence the way a worker's body develops, its optimal strategy to exploit that power is different.

This means there is a conflict of interests down on the farm. The queen is 'trying' to invest equally in males and females. The workers are trying to shift the ratio of reproductives in the direction of three females to every one male. If we are right to picture the workers as the farmers and the queen as their brood mare, presumably the workers will be successful in achieving their 3 : 1 ratio. If not, if the queen really lives up to her name and the workers are her slaves and the obedient tenders of the royal nurseries, then we should expect the 1 : 1 ratio which the queen 'prefers' to prevail. Who wins in this special case of a battle of the generations? This is a matter which can be put to the test and that is what Trivers and Hare did, using a large number of species of ants.

The sex ratio which is of interest is the ratio of male to female reproductives. These are the large winged forms which emerge from the ants' nest in periodic bursts for mating flights, after which the young queens may try to found new colonies. It is these winged forms which have to be counted to obtain an estimate of the sex ratio. Now the male and female reproductives are, in many species, very unequal in size. This complicates things since, as we saw in the previous chapter, the Fisher calculations about optimal sex ratio strictly apply, not to *numbers* of males and females, but to *quantity of investment* in males and females. Trivers and Hare made allowance for this by weighing them. They took 20 species of ant and estimated the sex ratio in terms of investment in reproductives. They found a rather convincingly close fit to the 3 : 1 female to male ratio predicted by the theory that the workers are running the show for their own benefit.

It seems then that in the ants studied, the conflict of interests is 'won' by the workers. This is not too surprising since worker bodies, being the guardians of the nurseries, have more power in practical terms than queen bodies. Genes trying to manipulate the world through queen bodies are outmanoeuvred by genes manipulating the world through worker bodies. It is interesting to look around for some special circumstances in which we might expect queens to have more practical power than workers. Trivers and Hare realized that there was just such a circumstance which could be used as a critical test of the theory.

This arises from the fact that there are some species of ant which take slaves. The workers of a slave-making species either do no ordinary work at all or are rather bad at it. What they are good at is going on slaving raids. True warfare in which large rival armies fight to the death is known only in man and in social insects. In many species of ants the specialized caste of workers known as soldiers have formidable fighting jaws, and devote their time to fighting for the colony against other ant armies. Slaving raids are just a particular kind of war effort. The slavers mount an attack on a nest of ants belonging to a different species, attempt to kill the defending workers or soldiers, and carry off the unhatched young. These young ones hatch out in the nest of their captors. They do not 'realize' that they are slaves and they set to work following their built-in nervous programs, doing all the duties that they would normally perform in their own nest. The slave-making workers or soldiers go on further slaving expeditions while the slaves stay at home and get on with the everyday business of running an ants' nest, cleaning, foraging, and caring for the brood.

The slaves are, of course, blissfully ignorant of the fact that they are unrelated to the queen and to the brood that they are tending. Unwittingly they are rearing new platoons of slave-makers. No doubt natural selection, acting on the genes of the slave species, tends to favour anti-slavery adaptations. However, these are evidently not fully effective because slavery is a wide-spread phenomenon.

The consequence of slavery which is interesting from our present point of view is this. The queen of the slave-making species is now in a position to bend the sex ratio in the direction she 'prefers'. This is because her own true-born children, the slavers, no longer hold the practical power in the nurseries. This power is now held by the slaves. The slaves 'think' they are looking after their own siblings and they are presumably doing whatever *would be appropriate in their own nests* to achieve their desired 3 : 1 bias in favour of sisters. But the queen of the slave-making species is able to get away with counter-measures and there is no selection operating on the slaves to neutralize these counter-measures, since the slaves are totally unrelated to the brood.

For example, suppose that in any ant species, queens 'attempt' to disguise male eggs by making them smell like female ones. Natural selection will normally favour any tendency by workers to 'see through' the disguise. We may picture an evolutionary battle in which queens continually 'change the code', and workers 'break the code'. The war will be won by whoever manages to get more of her genes into the next generation, via the bodies of the reproductives. This will normally be the workers, as we have seen. But when the queen of a *slave-making* species changes the code, the slave workers cannot evolve any ability to break the code. This is because any gene in a slave worker 'for breaking the code' is not represented in the body of any reproductive individual, and so is not passed on. The reproductives all belong to the slave-making species, and are kin to the queen but not to the slaves. If the genes of the slaves find their way into any reproductives at all, it will be into the reproductives which emerge from the original nest from which they were kidnapped. The slave workers will, if anything, be busy breaking the wrong code! Therefore, queens of a slave-making species can get away with changing their code freely, without there being any danger that genes for breaking the code will be propagated into the next generation.

The upshot of this involved argument is that we should expect in slave-making species that the ratio of investment in reproductives of the two sexes should approach $1 : 1$ rather than $3 : 1$. For once, the queen will have it all her own way. This is just what Trivers and Hare found, although they only looked at two slave-making species.

I must stress that I have told the story in an idealized way. Real life is not so neat and tidy. For instance, the most familiar social insect species of all, the honey bee, seems to do entirely the 'wrong' thing. There is a large surplus of investment in males over queens—something which does not appear to make sense from either the workers' or the mother queen's point of view. Hamilton has offered a possible solution to this puzzle. He points out that when a queen bee leaves the hive she goes with a large swarm of attendant workers, who help her to start a new colony. These workers are lost to the parent hive, and the cost of making them must be reckoned as part of the cost of reproduction: for every queen who leaves, many *extra* workers have to be made. Investment in these extra workers should be counted as part of the investment in reproductive females. The extra workers should be weighed in the balance against the males when the sex ratio is computed. So this was not a serious difficulty for the theory after all.

A more awkward spanner in the elegant works of the theory is the fact that, in some species, the young queen on her mating flight mates with several males instead of one. This means that the average relatedness among her daughters is less than $\frac{3}{4}$, and may even approach $\frac{1}{4}$ in extreme cases. It is tempting, though probably not very logical, to regard this as a cunning blow struck by queens against workers! Incidentally, this might seem to suggest that workers should chaperone a queen on her mating flight, to prevent her from mating more than once. But this would in no way help the workers' own genes—only the genes of the coming generation of workers. There is no trade-union spirit among the workers as a class. All that each one of them 'cares' about is her own genes. A worker might have 'liked' to have chaperoned her own mother, but she lacked the opportunity, not having been conceived in those days. A young queen on her mating flight is the sister of the present generation of workers, not the mother. Therefore they are on *her* side rather than on the side of the next generation of workers, who are merely their nieces. My head is now spinning, and it is high time to bring this topic to a close.

I have used the analogy of farming for what hymenopteran workers do to their mothers. The farm is a gene farm. The workers use their mother as a more efficient manufacturer of copies of their own genes than they would be themselves. The genes come off the production line in packages called reproductive individuals. This farming analogy should not be confused with a quite different sense in which the social insects may be said to farm. Social insects discovered, as man did long after, that settled cultivation of food can be more efficient than hunting and gathering.

For example, several species of ants in the New World, and, quite independently, termites in Africa, cultivate 'fungus gardens'. The best known are the so-called parasol ants of South America. These are immensely successful. Single colonies with more than two million individuals have been found. Their nests consist of huge spreading underground complexes of passages and galleries going down to a depth of ten feet or more, made by the excavation of as much as 40 tons of soil. The underground chambers contain the fungus gardens. The ants deliberately sow fungus of a particular species in special compost beds which they prepare by chewing leaves into fragments. Instead of foraging directly for their own food, the workers forage for leaves to make compost. The 'appetite' of a colony of parasol ants for leaves is gargantuan. This makes them a major economic pest, but the leaves are not food for themselves but food for their fungi. The ants eventually harvest and eat the fungi and feed them to their brood. The fungi are more efficient at breaking down leaf material than the ants' own stomachs would be, which is how the ants benefit by the arrangement. It is possible that the fungi benefit too, even though they are cropped: the ants propagate them more efficiently than their own spore dispersal mechanism might achieve. Furthermore, the ants 'weed' the fungus gardens, keeping them clear of alien species of fungi. By removing competition, this may benefit the ants' own domestic fungi. A kind of relationship of mutual altruism could be said to exist between ants and fungi. It is remarkable that a very similar system of fungus-farming has evolved independently, among the quite unrelated termites.

Ants have their own domestic animals as well as their crop plants. Aphids—greenfly and similar bugs—are highly specialized for sucking the juice out of plants. They pump the sap up out of the plants' veins more efficiently than they subsequently digest it. The result is that they excrete a liquid which has had only some of its nutritious value extracted. Droplets of sugar-rich 'honeydew' pass out of the back end at a great rate, in some cases more than the insect's own body-weight every hour. The honeydew normally rains down on to the ground—it may well have been the providential food known as 'manna' in the Old Testament. But ants of several species intercept it as soon as it leaves the bug. The ants 'milk' the aphids by stroking their hind-quarters with their feelers and legs. Aphids respond to this, in some cases apparently holding back their droplets until an ant strokes them, and even withdrawing a droplet if an ant is not ready to accept it. It has been suggested that some aphids have evolved a backside which looks and feels like an ant's face, the better to attract ants. What the aphids have to gain from the relationship is apparently protection from their natural enemies. Like our own dairy cattle they lead a sheltered life, and aphid species which are much cultivated by ants have lost their normal defensive mechanisms. In some cases ants care for the aphid eggs inside their own underground nests, feed the young aphids, and finally, when they are grown, gently carry them up to the protected grazing grounds.

A relationship of mutual benefit between members of different species is called mutualism or symbiosis. Members of different species often have much to offer each other because they can bring different 'skills' to the partnership. This kind of fundamental asymmetry can lead to evolutionarily stable strategies of mutual cooperation. Aphids have the right sort of mouthparts for pumping up plant sap, but such sucking mouthparts are no good for self-defence. Ants are no good at sucking sap from plants, but they are good at fighting. Ant genes for cultivating and protecting aphids have been favoured in ant gene-pools. Aphid genes for cooperating with the ants have been favoured in aphid gene-pools.

Symbiotic relationships of mutual benefit are common among animals and plants. A lichen appears superficially to be an individual plant like any other. But it is really an intimate symbiotic union between a fungus and a green alga. Neither partner could live without the other. If their union had become just a bit more intimate we would no longer have been able to tell that a lichen was a double organism at all. Perhaps then there are other double or multiple organisms which we have not recognized as such. Perhaps even we ourselves?

Within each one of our cells there are numerous tiny bodies called mitochondria. The mitochondria are chemical factories, responsible for providing most of the energy we need. If we lost our mitochondria we would be dead within seconds. Recently it has been plausibly argued that mitochondria are, in origin, symbiotic bacteria who joined forces with our type of cell very early in evolution. Similar suggestions have been made for other small bodies within our cells. This is one of those revolutionary ideas which it takes time to get used to, but it is an idea whose time has come. I speculate that we shall come to accept the more radical idea that each one of our genes is a symbiotic unit. We are gigantic colonies of symbiotic genes. One cannot really speak of 'evidence' for this idea, but, as I tried to suggest in earlier chapters, it is really inherent in the very way we think about how genes work in sexual species. The other side of this coin is that viruses may be genes who have broken loose from 'colonies' such as ourselves. Viruses consist of pure DNA (or a related self-replicating molecule) surrounded by a protein jacket. They are all parasitic. The suggestion is that they have evolved from 'rebel' genes who escaped, and now travel from body to body directly through the air, rather than via the more conventional vehicles—sperms and eggs. If this is true, we might just as well regard ourselves as colonies of viruses! Some of them cooperate symbiotically, and travel from body to body in sperms and eggs. These are the conventional 'genes'. Others live parasitically, and travel by whatever means they can. If the parasitic DNA travels in sperms and eggs, it perhaps forms the 'paradoxical' surplus of

DNA which I mentioned in Chapter 3. If it travels through the air, or by other direct means, it is called 'virus' in the usual sense.

But these are speculations for the future. At present we are concerned with symbiosis at the higher level of relationships between many-celled organisms, rather than within them. The word symbiosis is conventionally used for associations between members of different species. But, now that we have eschewed the 'good of the species' view of evolution, there seems no logical reason to distinguish associations between members of different species as things apart from associations between members of the same species. In general, associations of mutual benefit will evolve if each partner can get more out than he puts in. This is true whether we are speaking of members of the same hyena pack, or of widely distinct creatures such as ants and aphids, or bees and flowers. In practice it may be difficult to distinguish cases of genuine two-way mutual benefit from cases of one-sided exploitation.

The evolution of associations of mutual benefit is theoretically easy to imagine if the favours are given and received simultaneously, as in the case of the partners who make up a lichen. But problems arise if there is a delay between the giving of a favour and its repayment. This is because the first recipient of a favour may be tempted to cheat and refuse to pay it back when his turn comes. The resolution of this problem is interesting and is worth discussing in detail. I can do this best in terms of a hypothetical example.

Suppose a species of bird is parasitized by a particularly nasty kind of tick which carries a dangerous disease. It is very important that these ticks should be removed as soon as possible. Normally an individual bird can pull off its own ticks when preening itself. There is one place, however—the top of the head—which it cannot reach with its own bill. The solution to the problem quickly occurs to any human. An individual may not be able to reach his own head, but nothing is easier than for a friend to do it for him. Later, when the friend is parasitized himself, the good deed can be paid back. Mutual grooming is in fact very common in both birds and mammals.

This makes immediate intuitive sense. Anybody with conscious foresight can see that it is sensible to enter into mutual back-scratching arrangements. But we have learnt to beware of what seems intuitively sensible. The gene has no foresight. Can the theory of selfish genes account for mutual back-scratching, or 'reciprocal altruism', where there is a delay between good deed and repayment? Williams briefly discussed the problem in his 1966 book, to which I have already referred. He concluded, as had Darwin, that delayed reciprocal altruism can evolve in species which are capable of recognizing and remembering each other as individuals. Trivers, in 1971, took the matter further. When he wrote, he did not have available to him Maynard Smith's concept of the evolutionarily stable strategy. If he had, my guess is that he would have made use of it, for it provides a natural way to express his ideas. His reference to the 'prisoner's dilemma'—a favourite puzzle in Game Theory—shows that he was already thinking along the same lines.

Suppose *B* has a parasite on the top of his head. *A* pulls it off him. Later, the time comes when *A* has a parasite on his head. He naturally seeks out *B* in order that *B* may pay back his good deed. *B* simply turns up his nose and walks off. *B* is a cheat, an individual who accepts the benefit of other individuals' altruism, but who does not pay it back, or who pays it back insufficiently. Cheats do better than indiscriminate altruists because they gain the benefits without paying the costs. To be sure, the cost of grooming another individual's head seems small compared with the benefit of having a dangerous parasite removed, but it is not negligible. Some valuable energy and time has to be spent.

Let the population consist of individuals who adopt one of two strategies. As in Maynard Smith's analyses, we are not talking about conscious strategies, but about unconscious behaviour programs laid down by genes. Call the two strategies Sucker and Cheat. Suckers groom anybody who needs it, indiscriminately. Cheats accept altruism from suckers, but they never groom anybody else, not even somebody who has previously groomed them. As in the case of the hawks and doves, we arbitrarily assign pay-off points. It does not matter what the exact values are, so long as the benefit of being groomed exceeds the cost of grooming. If the incidence of parasites is high, any individual sucker in a population of suckers can reckon on being groomed about as often as he grooms. The average pay-off for a sucker among suckers is therefore positive. They all do quite nicely in fact, and the word sucker seems inappropriate. But now suppose a cheat arises in the population. Being the only cheat, he can count on being groomed by everybody else, but he pays nothing in return. His average pay-off is better than the average for a sucker. Cheat genes will therefore start to spread through the population. Sucker genes will soon be driven to extinction. This is because, no matter what the ratio in the population, cheats will always do better than suckers. For instance, consider the case when the population consists of 50 per cent suckers and 50 per cent cheats. The average pay-off for both suckers and cheats will be less than that for any individual in a population of 100 per cent suckers. But still, cheats will be doing better than suckers because they are getting all the benefits—such as they are—and paying nothing back. When the proportion of cheats reaches 90 per cent, the average pay-off for all individuals will be very low: many of both types may by now be dying of the infection carried by the ticks. But still the cheats will be doing better than the suckers. Even if the whole population declines toward extinction, there will never be any time when suckers do better than cheats. Therefore, as long as we consider only these two strategies, nothing can stop the extinction of the suckers and, very probably, the extinction of the whole population too.

But now, suppose there is a third strategy called Grudger. Grudgers groom strangers and individuals who have previously groomed them. However, if any individual cheats them, they remember the incident and bear a grudge: they refuse to groom that individual in the future. In a population of grudgers and suckers it is impossible to tell which is which. Both types behave altruistically towards everybody else, and both earn an equal and high average pay-off. In a population consisting largely of cheats, a single grudger would not be very successful. He would expend a great deal of energy grooming most of the individuals he met—for it would take time for him to build up grudges against all of them. On the other hand, nobody would groom him in return. If grudgers are rare in comparison with cheats, the grudger gene will go extinct. Once the grudgers manage to build up in numbers so that they reach a critical proportion, however, their chance of meeting each other becomes sufficiently great to off-set their wasted effort in grooming cheats. When this critical proportion is reached they will start to average a higher pay-off than cheats, and the cheats will be driven at an accelerating rate towards extinction. When the cheats are nearly extinct their rate of decline will become slower, and they may survive as a minority for quite a long time. This is because for any one rare cheat there is only a small chance of his encountering the same grudger twice: therefore the proportion of individuals in the population who bear a grudge against any given cheat will be small.

I have told the story of these strategies as though it were intuitively obvious what would happen. In fact it is not all that obvious, and I did take the precaution of simulating it on a computer to check that intuition was right. Grudger does indeed turn out to be an evolutionarily stable strategy against sucker and cheat, in the sense that, in a population consisting largely of grudgers, neither cheat nor sucker will invade. Cheat is also an ESS, however, because a population consisting largely of cheats will not be invaded by either grudger or sucker. A population could sit at either of these two ESSs. In the long term it might flip from one to the other. Depending on the exact values of the pay-offs—the assumptions in the simulation were of course completely arbitrary—one or other of the two stable states will have a larger 'zone of attraction' and will be more likely to be attained. Note incidentally that, although a population of cheats may be more likely to go extinct than a population of grudgers, this in no way affects its status as an ESS. If a population arrives at an ESS which drives it extinct, then it goes extinct, and that is just too bad.

It is quite entertaining to watch a computer simulation which starts with a strong majority of suckers, a minority of grudgers which is just above the critical frequency, and about the same-sized minority of cheats. The first thing that happens is a dramatic crash in the population of suckers as the cheats ruthlessly exploit them. The cheats enjoy a soaring population explosion, reaching their peak just as the last sucker perishes. But the cheats still have the grudgers to reckon with. During the precipitous decline of the suckers, the grudgers have been slowly decreasing in numbers, taking a battering from the prospering cheats, but just managing to hold their own. After the last sucker has gone and the cheats can no longer get away with selfish exploitation so easily, the grudgers slowly begin to increase at the cheats' expense. Steadily their population rise gathers momentum. It accelerates steeply, the cheat population crashes to near extinction, then levels out as they enjoy the privileges of rarity and the comparative freedom from grudges which this brings. However, slowly and inexorably the cheats are driven out of existence, and the grudgers are left in sole possession. Paradoxically, the presence of the suckers actually endangered the grudgers early on in the story because they were responsible for the temporary prosperity of the cheats.

By the way, my hypothetical example about the dangers of not being groomed is quite plausible. Mice kept in isolation tend to develop unpleasant sores on those parts of their heads which they cannot reach. In one study, mice kept in groups did not suffer in this way, because they licked each others' heads. It would be interesting to test the theory of reciprocal altruism experimentally and it seems that mice might be suitable subjects for the work.

Trivers discusses the remarkable symbiosis of the cleaner-fish. Some fifty species, including small fish and shrimps, are known to make their living by picking parasites off the surface of larger fish of other species. The large fish obviously benefit from being cleaned, and the cleaners get a good supply of food. The relationship is symbiotic. In many cases the large fish open their mouths and allow cleaners right inside to pick their teeth, and then to swim out through the gills which they also clean. One might expect that a large fish would craftily wait until he had been thoroughly cleaned, and then gobble up the cleaner. Yet instead he usually lets the cleaner swim off unmolested. This is a considerable feat of apparent altruism because in many cases the cleaner is of the same size as the large fish's normal prey.

Cleaner-fish have special stripy patterns and special dancing displays which label them as cleaners. Large fish tend to refrain from eating small fish who have the right kind of stripes, and who approach them with the right kind of dance. Instead they go into a trance-like state and allow the cleaner free access to their exterior and interior. Selfish genes being what they are, it is not surprising that ruthless, exploiting cheats have cashed in. There are species of small fish that look just like cleaners and dance in the same kind of way in order to secure safe conduct into the vicinity of large fish. When the large fish has gone into its expectant trance the cheat, instead of pulling off a parasite, bites a chunk out of the large fish's fin and beats a hasty retreat. But in spite of the cheats, the relationship between fish cleaners and their clients is mainly amicable and stable. The profession of cleaner plays an important part in the daily life of the coral reef community. Each cleaner has his own territory, and large fish have been seen queuing up for attention like customers at a barber's shop. It is probably this site-tenacity which makes possible the evolution of delayed reciprocal-altruism in this case. The benefit to a large fish of being able to return repeatedly to the same 'barber's shop', rather than continually searching for a new one, must outweigh the cost of refraining from eating the cleaner. Since cleaners are small, this is not hard to believe. The presence of cheating cleaner-mimics probably indirectly endangers the bona-fide cleaners by setting up a minor pressure on large fish to eat stripy dancers. Site-tenacity on the part of genuine cleaners enables customers to find them and to avoid cheats.

A long memory and a capacity for individual recognition are well developed in man. We might therefore expect reciprocal altruism to have played an important part in human evolution. Trivers goes so far as to suggest that many of our psychological characteristics—envy, guilt, gratitude, sympathy etc.—have been shaped by natural selection for improved ability to cheat, to detect cheats, and to avoid being thought to be a cheat. Of particular interest are 'subtle cheats' who appear to be reciprocating, but who consistently pay back slightly less than they receive. It is even possible that man's swollen brain, and his predisposition to reason mathematically, evolved as a mechanism of ever more devious cheating, and ever more penetrating detection of cheating in others. Money is a formal token of delayed reciprocal altruism.

There is no end to the fascinating speculation which the idea of reciprocal altruism engenders when we apply it to our own species. Tempting as it is, I am no better at such speculation than the next man, and I leave the reader to entertain himself.

11. Memes: the new replicators

So far, I have not talked much about man in particular, though I have not deliberately excluded him either. Part of the reason I have used the term 'survival machine' is that 'animal' would have left out plants and, in some people's minds, humans. The arguments I have put forward should, prima facie, apply to any evolved being. If a species is to be excepted, it must be for good particular reasons. Are there any good reasons for supposing our own species to be unique? I believe the answer is yes.

Most of what is unusual about man can be summed up in one word 'culture'. I use the word not in its snobbish sense, but as a scientist uses it. Cultural transmission is analogous to genetic transmission in that, although basically conservative, it can give rise to a form of evolution. Geoffrey Chaucer could not hold a conversation with a modern Englishman, even though they are linked to each other by an unbroken chain of some twenty generations of Englishmen, each of whom could speak to his immediate neighbours in the chain as a son speaks to his father. Language seems to 'evolve' by non-genetic means, and at a rate which is orders of magnitude faster than genetic evolution.

Cultural transmission is not unique to man. The best non-human example that I know has recently been described by P. F. Jenkins in the song of a bird called the saddleback which lives on islands off New Zealand. On the island where he worked there was a total repertoire of about nine distinct songs. Any given male sang only one or a few of these songs. The males could be classified into dialect groups. For example, one group of eight males with neighbouring territories sang a particular song called the CC song. Other dialect groups sang different songs. Sometimes the members of a dialect group shared more than one distinct song. By comparing the songs of fathers and sons, Jenkins showed that song patterns were not inherited genetically. Each young male was likely to adopt songs from his territorial neighbours by imitation, in an analogous way to human language. During most of the time Jenkins was there, there was a fixed number of songs on the island, a kind of 'song pool' from which each young male drew his own small repertoire. But occasionally Jenkins was privileged to witness the 'invention' of a new song, which occurred by a mistake in the imitation of an old one. He writes: 'New song forms have been shown to arise variously by change of pitch of a note, repetition of a note, the elision of notes and the combination of parts of other existing songs . . . The appearance of the new form was an abrupt event and the product was quite stable over a period of years. Further, in a number of cases the variant was transmitted accurately in its new form to younger recruits so that a recognizably coherent group of like singers developed.' Jenkins refers to the origins of new songs as 'cultural mutations'.

Song in the saddleback truly evolves by non-genetic means. There are other examples of cultural evolution in birds and monkeys, but these are just interesting oddities. It is our own species that really shows what cultural evolution can do. Language is only one example out of many. Fashions in dress and diet, ceremonies and customs, art and architecture, engineering and technology, all evolve in historical time in a way that looks like highly speeded up genetic evolution, but has really nothing to do with genetic evolution. As in genetic evolution though, the change may be progressive. There is a sense in which modern science is actually better than ancient science. Not only does our understanding of the universe change as the centuries go by: it improves. Admittedly the current burst of improvement dates back only to the Renaissance, which was preceded by a dismal period of stagnation, in which European scientific culture was frozen at the level achieved by the Greeks. But, as we saw in Chapter 5, genetic evolution too may proceed as a series of brief spurts between stable plateaux.

The analogy between cultural and genetic evolution has frequently been pointed out, sometimes in the context of quite unnecessary mystical overtones. The analogy between scientific progress and genetic evolution by natural selection has been illuminated especially by Sir Karl Popper. I want to go even further into directions which are also being explored by, for example, the geneticist L. L. Cavalli-Sforza, the anthropologist F. T. Cloak, and the ethologist J. M. Cullen.

As an enthusiastic Darwinian, I have been dissatisfied with explanations which my fellow-enthusiasts have offered for human behaviour. They have tried to look for 'biological advantages' in various attributes of human civilization. For instance, tribal religion has been seen as a mechanism for solidifying group identity, valuable for a pack-hunting species whose individuals rely on cooperation to catch large and fast prey. Frequently the evolutionary preconception in terms of which such theories are framed is implicitly group-selectionist, but it is possible to rephrase the theories in terms of orthodox gene selection. Man may well have spent large portions of the last several million years living in small kin groups. Kin selection and selection in favour of reciprocal altruism may have acted on human genes to produce many of our basic psychological attributes and tendencies. These ideas are plausible as far as they go, but I find that they do not begin to square up to the formidable challenge of explaining culture, cultural evolution, and the immense differences between human cultures around the world, from the utter selfishness of the Ik of Uganda, as described by Colin Turnbull, to the gentle altruism of Margaret Mead's Arapesh. I think we have got to start again and go right back to first principles. The argument I shall advance, surprising as it may seem coming from the author of the earlier chapters, is that, for an understanding of the evolution of modern man, we must begin by throwing out the gene as the sole basis of our ideas on evolution. I am an enthusiastic Darwinian, but I think Darwinism is too big a theory to be confined to the narrow context of the gene. The gene will enter my thesis as an analogy, nothing more.

What, after all, is so special about genes? The answer is that they are replicators. The laws of physics are supposed to be true all over the accessible universe. Are there any principles of biology which are likely to have similar universal validity? When astronauts voyage to distant planets and look for life, they can expect to find creatures too strange and unearthly for us to imagine. But is there anything which must be true of all life, wherever it is found, and whatever the basis of its chemistry? If forms of life exist whose chemistry is based on silicon rather than carbon, or ammonia rather than water, if creatures are discovered which boil to death at — 100 degrees centigrade, if a form of life is found which is not based on chemistry at all but on electronic reverberating circuits, will there still be any general principle which is true of all life? Obviously I do not know but, if I had to bet, I would put my money on one fundamental principle. This is the law that all life evolves by the differential survival of replicating entities. The gene, the DNA molecule, happens to be the replicating entity which prevails on our own planet. There may be others. If there are, provided certain other conditions are met, they will almost inevitably tend to become the basis for an evolutionary process.

But do we have to go to distant worlds to find other kinds of replicator and other, consequent, kinds of evolution? I think that a new kind of replicator has recently emerged on this very planet. It is staring us in the face. It is still in its infancy, still drifting clumsily about in its primeval soup, but already it is achieving evolutionary change at a rate which leaves the old gene panting far behind.

The new soup is the soup of human culture. We need a name for the new replicator, a noun which conveys the idea of a unit of cultural transmission, or a unit of *imitation*. 'Mimeme' comes from a suitable Greek root, but I want a monosyllable that sounds a bit like 'gene'. I hope my classicist friends will forgive me if I abbreviate mimeme to *meme*. If it is any consolation, it could alternatively be thought of as being related to 'memory', or to the French word *même*. It should be pronounced to rhyme with 'cream'.

Examples of memes are tunes, ideas, catch-phrases, clothes fashions, ways of making pots or of building arches. Just as genes propagate themselves in the gene pool by leaping from body to body via sperms or eggs, so memes propagate themselves in the meme pool by leaping from brain to brain via a process which, in the broad sense, can be called imitation. If a scientist hears, or reads about, a good idea, he passes it on to his colleagues and students. He mentions it in his articles and his lectures. If the idea catches on, it can be said to propagate itself, spreading from brain to brain. As my colleague N. K. Humphrey neatly summed up an earlier draft of this chapter: '. . . memes should be regarded as living structures, not just metaphorically but technically. When you plant a fertile meme in my mind you literally parasitize my brain, turning it into a vehicle for the meme's propagation in just the way that a virus may parasitize the genetic mechanism of a host cell. And this isn't just a way of talking—the meme for, say, 'belief in life after death' is actually realized physically, millions of times over, as a structure in the nervous systems of individual men the world over.'

Consider the idea of God. We do not know how it arose in the meme pool. Probably it originated many times by independent 'mutation'. In any case, it is very old indeed. How does it replicate itself? By the spoken and written word, aided by great music and great art. Why does it have such high survival value? Remember that 'survival value' here does not mean value for a gene in a gene pool, but value for a meme in a meme pool. The question really means: What is it about the idea of a god which gives it its stability and penetrance in the cultural environment? The survival value of the god meme in the meme pool results from its great psychological appeal. It provides a superficially plausible answer to deep and troubling questions about existence. It suggests that injustices in this world may be rectified in the next. The 'everlasting arms' hold out a cushion against our own inadequacies which, like a doctor's placebo, is none the less effective for being imaginary. These are some of the reasons why the idea of God is copied so readily by successive generations of individual brains. God exists, if only in the form of a meme with high survival value, or infective power, in the environment provided by human culture.

Some of my colleagues have suggested to me that this account of the survival value of the god meme begs the question. In the last analysis they wish always to go back to 'biological advantage'. To them it is not good enough to say that the idea of a god has 'great psychological appeal'. They want to know *why* it has great psychological appeal. Psychological appeal means appeal to brains, and brains are shaped by natural selection of genes in gene-pools. They want to find some way in which having a brain like that improves gene survival.

I have a lot of sympathy with this attitude, and I do not doubt that there are genetic advantages in our having brains of the kind which we have. But nevertheless I think that these colleagues, if they look carefully at the fundamentals of their own assumptions, will find that they are begging just as many questions as I am. Fundamentally, the reason why it is good policy for us to try to explain biological phenomena in terms of gene advantage is that genes are replicators. As soon as the primeval soup provided conditions in which molecules could make copies of themselves, the replicators themselves took over. For more than three thousand million years, DNA has been the only replicator worth talking about in the world. But it does not necessarily hold these monopoly rights for all time. Whenever conditions arise in which a new kind of replicator *can* make copies of itself, the new replicators *will* tend to take over, and start a new kind of evolution of their own. Once this new evolution begins, it will in no necessary sense be subservient to the old. The old gene-selected evolution, by making brains, provided the 'soup' in which the first memes arose. Once self-copying memes had arisen, their own, much faster, kind of evolution took off. We biologists have assimilated the idea of genetic evolution so deeply that we tend to forget that it is only one of many possible kinds of evolution.

Imitation, in the broad sense, is how memes *can* replicate. But just as not all genes which can replicate do so successfully, so some memes are more successful in the meme-pool than others. This is the analogue of natural selection. I have mentioned particular examples of qualities which make for high survival value among memes. But in general they must be the same as those discussed for the replicators of Chapter 2: longevity, fecundity, and copying-fidelity. The longevity of any one copy of a meme is probably relatively unimportant, as it is for any one copy of a gene. The copy of the tune 'Auld Lang Syne' which exists in my brain will last only for the rest of my life. The copy of the same tune which is printed in my volume of *The Scottish Student's Song Book* is unlikely to last much longer. But I expect there will be copies of the same tune on paper and in peoples' brains for centuries to come. As in the case of genes, fecundity is much more important than longevity of particular copies. If the meme is a scientific idea, its spread will depend on how acceptable it is to the population of individual scientists; a rough measure of its survival value could be obtained by counting the number of times it is referred to in successive years in scientific journals. If it is a popular tune, its spread through the meme pool may be gauged by the number of people heard whistling it in the streets. If it is a style of women's shoe, the population memeticist may use sales statistics from shoe shops. Some memes, like some genes, achieve brilliant short-term success in spreading rapidly, but do not last long in the meme pool. Popular songs and stiletto heels are examples. Others, such as the Jewish religious laws, may continue to propagate themselves for thousands of years, usually because of the great potential permanence of written records.

This brings me to the third general quality of successful replicators: copying-fidelity. Here I must admit that I am on shaky ground. At first sight it looks as if memes are not high-fidelity replicators at all. Every time a scientist hears an idea and passes it on to somebody else, he is likely to change it somewhat. I have made no secret of my debt in this book to the ideas of R. L. Trivers. Yet I have not repeated them in his own words. I have twisted them round for my own purposes, changing the emphasis, blending them with ideas of my own and of other people. The memes are being passed on to you in altered form. This looks quite unlike the particulate, all-or-none quality of gene transmission. It looks as though meme transmission is subject to continuous mutation, and also to blending.

It is possible that this appearance of non-particulateness is illusory, and that the analogy with genes does not break down. After all, if we look at the inheritance of many genetic characters such as human height or skin-colouring, it does not look like the work of indivisible and unblendable genes. If a black and a white person mate, their children do not come out either black or white: they are intermediate. This does not mean the genes concerned are not particulate. It is just that there are so many of them concerned with skin colour, each one having such a small effect, that they *seem* to blend. So far I have talked of memes as though it was obvious what a single unit-meme consisted of. But of course it is far from obvious. I have said a tune is one meme, but what about a symphony: how many memes is that? Is each movement one meme, each recognizable phrase of melody, each bar, each chord, or what?

I appeal to the same verbal trick as I used in Chapter 3. There

I divided the 'gene complex' into large and small genetic units, and units within units. The 'gene' was defined, not in a rigid all-or-none way, but as a unit of convenience, a length of chromosome with just sufficient copying-fidelity to serve as a viable unit of natural selection. If a single phrase of Beethoven's ninth symphony is sufficiently distinctive and memorable to be abstracted from the context of the whole symphony, and used as the call-sign of a maddeningly intrusive European broadcasting station, then to that extent it deserves to be called one meme. It has, incidentally, materially diminished my capacity to enjoy the original symphony.

Similarly, when we say that all biologists nowadays believe in Darwin's theory, we do not mean that every biologist has, graven in his brain, an identical copy of the exact words of Charles Darwin himself. Each individual has his own way of interpreting Darwin's ideas. He probably learned them not from Darwin's own writings, but from more recent authors. Much of what Darwin said is, in detail, wrong. Darwin if he read this book would scarcely recognize his own original theory in it, though I hope he would like the way I put it. Yet, in spite of all this, there is something, some essence of Darwinism, which is present in the head of every individual who understands the theory. If this were not so, then almost any statement about two people agreeing with each other would be meaningless. An 'idea-meme' might be defined as an entity which is capable of being transmitted from one brain to another. The meme of Darwin's theory is therefore that essential basis of the idea which is held in common by all brains who understand the theory. The *differences* in the ways that people represent the theory are then, by definition, not part of the meme. If Darwin's theory can be subdivided into compon-ents, such that some people believe component A but not com-ponent B, while others believe B but not A, then A and B should be regarded as separate memes. If almost everybody who believes in A also believes in B—if the memes are closely 'linked' to use the genetic term—then it is convenient to lump them together as one meme.

Let us pursue the analogy between memes and genes further. Throughout this book, I have emphasized that we must not think of genes as conscious, purposeful agents. Blind natural selection, however, makes them behave rather as if they were purposeful, and it has been convenient, as a shorthand, to refer to genes in the language of purpose. For example, when we say 'genes are trying to increase their numbers in future gene pools', what we really mean is 'those genes which behave in such a way as to increase their numbers in future gene pools tend to be the genes whose effects we see in the world'. Just as we have found it convenient to think of genes as active agents, working purposefully for their own survival, perhaps it might be convenient to think of memes in the same way. In neither case must we get mystical about it. In both cases the idea of purpose is only a metaphor, but we have already seen what a fruitful metaphor it is in the case of genes. We have even used words like 'selfish' and 'ruthless' of genes, knowing full well it is only a figure of speech. Can we, in exactly the same spirit, look for selfish or ruthless memes?

There is a problem here concerning the nature of competition. Where there is sexual reproduction, each gene is competing particularly with its own alleles—rivals for the same chromosomal slot. Memes seem to have nothing equivalent to chromosomes, and nothing equivalent to alleles. I suppose there is a trivial sense in which many ideas can be said to have 'opposites'. But in general memes resemble the early replicating molecules, floating chaotically free in the primeval soup, rather than modern genes in their neatly paired, chromosomal regiments. In what sense then are memes competing with each other? Should we expect them to be 'selfish' or 'ruthless', if they have no alleles? The answer is that we might, because there is a sense in which they must indulge in a kind of competition with each other.

Any user of a digital computer knows how precious computer time and memory storage space are. At many large computer centres they are literally costed in money; or each user may be allotted a ration of time, measured in seconds, and a ration of space, measured in 'words'. The computers in which memes live are human brains. Time is possibly a more important limiting factor than storage space, and it is the subject of heavy competition. The human brain, and the body which it controls, cannot do more than one or a few things at once. If a meme is to dominate the attention of a human brain, it must do so at the expense of 'rival' memes. Other commodities for which memes compete are radio and television time, billboard space, newspaper column-inches, and library shelf-space.

In the case of genes, we saw in Chapter 3 that co-adapted gene complexes may arise in the gene pool. A large set of genes concerned with mimicry in butterflies became tightly linked together on the same chromosome, so tightly that they can be treated as one gene. In Chapter 5 we met the more sophisticated idea of the evolutionarily stable set of genes. Mutually suitable teeth, claws, guts, and sense organs evolved in carnivore gene pools, while a different stable set of characteristics emerged from herbivore gene pools. Does anything analogous occur in meme pools? Has the god meme, say, become associated with any other particular memes, and does this association assist the survival of each of the participating memes? Perhaps we could regard an organized church, with its architecture, rituals, laws, music, art, and written tradition, as a co-adapted stable set of mutually-assisting memes.

To take a particular example, an aspect of doctrine which has been very effective in enforcing religious observance is the threat of hell fire. Many children and even some adults believe that they will suffer ghastly torments after death if they do not obey the priestly rules. This is a peculiarly nasty technique of persuasion, causing great psychological anguish throughout the middle ages and even today. But it is highly effective. It might almost have been planned deliberately by a machiavellian priesthood trained in deep psychological indoctrination techniques. However, I doubt if the priests were that clever. Much more probably, unconscious memes have ensured their own survival by virtue of those same qualities of pseudo-ruthlessness which successful genes display. The idea of hell fire is, quite simply, *self perpetuating*, because of its own deep psychological impact. It has become linked with the god meme because the two reinforce each other, and assist each other's survival in the meme pool.

Another member of the religious meme complex is called faith. It means blind trust, in the absence of evidence, even in the teeth of evidence. The story of Doubting Thomas is told, not so that we shall admire Thomas, but so that we can admire the other apostles in comparison. Thomas demanded evidence. Nothing is more lethal for certain kinds of meme than a tendency to look for evidence. The other apostles, whose faith was so strong that they did not need evidence, are held up to us as worthy of imitation. The meme for blind faith secures its own perpetuation by the simple unconscious expedient of discouraging rational inquiry.

Blind faith can justify anything. If a man believes in a different god, or even if he uses a different ritual for worshipping the same god, blind faith can decree that he should die—on the cross, at the stake, skewered on a Crusader's sword, shot in a Beirut street, or blown up in a bar in Belfast. Memes for blind faith have their own ruthless ways of propagating themselves. This is true of patriotic and political as well as religious blind faith.

Memes and genes may often reinforce each other, but they sometimes come into opposition. For example, the habit of celibacy is presumably not inherited genetically. A gene for celibacy is doomed to failure in the gene pool, except under very special circumstances such as we find in the social insects. But still, a *meme* for celibacy can be successful in the meme pool. For example, suppose the success of a meme depends critically on how much time people spend in actively transmitting it to other people. Any time spent in doing other things than attempting to transmit the meme may be regarded as time wasted from the meme's point of view. The meme for celibacy is transmitted by priests to young boys who have not yet decided what they want to do with their lives. The medium of transmission is human influence of various kinds, the spoken and written word, personal example and so on. Suppose, for the sake of argument, it happened to be the case that marriage weakened the power of a priest to influence his flock, say because it occupied a large proportion of his time and attention. This has, indeed, been advanced as an official reason for the enforcement of celibacy among priests. If this were the case, it would follow that the meme for celibacy could have greater survival value than the meme for marriage. Of course, exactly the opposite would be true for a *gene* for celibacy. If a priest is a survival machine for memes, celibacy is a useful attribute to build into him. Celibacy is just a minor partner in a large complex of mutually-assisting religious memes.

I conjecture that co-adapted meme-complexes evolve in the same kind of way as co-adapted gene-complexes. Selection favours memes which exploit their cultural environment to their own advantage. This cultural environment consists of other memes which are also being selected. The meme pool therefore comes to have the attributes of an evolutionarily stable set, which new memes find it hard to invade.

I have been a bit negative about memes, but they have their cheerful side as well. When we die there are two things we can leave behind us: genes and memes. We were built as gene machines, created to pass on our genes. But that aspect of us will be forgotten in three generations. Your child, even your grandchild, may bear a resemblance to you, perhaps in facial features, in a talent for music, in the colour of her hair. But as each generation passes, the contribution of your genes is halved. It does not take long to reach negligible proportions. Our genes may be immortal but the *collection* of genes which is any one of us is bound to crumble away. Elizabeth II is a direct descendant of William the Conqueror. Yet it is quite probable that she bears not a single one of the old king's genes. We should not seek immortality in reproduction.

But if you contribute to the world's culture, if you have a good idea, compose a tune, invent a sparking plug, write a poem, it may live on, intact, long after your genes have dissolved in the common pool. Socrates may or may not have a gene or two alive in the world today, as G. C. Williams has remarked, but who cares? The meme-complexes of Socrates, Leonardo, Copernicus, and Marconi are still going strong.

However speculative my development of the theory of memes may be, there is one serious point which I would like to emphasize once again. This is that when we look at the evolution of cultural traits and at their survival value, we must be clear *whose* survival we are talking about. Biologists, as we have seen, are accustomed to looking for advantages at the gene level (or the individual, the group, or the species level according to taste). What we have not previously considered is that a cultural trait may have evolved in the way that it has, simply because it is *advantageous to itself*.

We do not have to look for conventional biological survival values of traits like religion, music, and ritual dancing, though these may also be present. Once the genes have provided their survival machines with brains which are capable of rapid imitation, the memes will automatically take over. We do not even have to posit a genetic advantage in imitation, though that would certainly help. All that is necessary is that the brain should be *capable* of imitation: memes will then evolve which exploit the capability to the full.

I now close the topic of the new replicators, and end the book on a note of qualified hope. One unique feature of man, which may or may not have evolved memically, is his capacity for conscious foresight. Selfish genes (and, if you allow the speculation of this chapter, memes too) have no foresight. They are unconscious, blind, replicators. The fact that they replicate, together with certain further conditions means, willy nilly, that they will tend towards the evolution of qualities which, in the special sense of this book, can be called selfish. A simple replicator, whether gene or meme, cannot be expected to forgo short-term selfish advantage even if it would really pay it, in the long term, to do so. We saw this in the chapter on aggression. Even though a 'conspiracy of doves' would be better for *every single individual* than the evolutionarily stable strategy, natural selection is bound to favour the ESS.

It is possible that yet another unique quality of man is a capacity for genuine, disinterested, true altruism. I hope so, but I am not going to argue the case one way or the other, nor to speculate over its possible memic evolution. The point I am making now is that, even if we look on the dark side and assume that individual man is fundamentally selfish, our conscious foresight—our capacity to simulate the future in imagination—could save us from the worst selfish excesses of the blind replicators. We have at least the mental equipment to foster our long-term selfish interests rather than merely our short-term selfish interests. We can see the long-term benefits of participating in a 'conspiracy of doves', and we can sit down together to discuss ways of making the conspiracy work. We have the power to defy the selfish genes of our birth and, if necessary, the selfish memes of our indoctrination. We can even discuss ways of deliberately cultivating and nurturing pure, disinterested altruism— something that has no place in nature, something that has never existed before in the whole history of the world. We are built as gene machines and cultured as meme machines, but we have the power to turn against our creators. We, alone on earth, can rebel against the tyranny of the selfish replicators.

Bibliography

Not all the works listed here are mentioned by name in the book, but all of them are referred to by number in the index.

1. ALEXANDER, R. D. (1961). Aggressiveness, territoriality, and sexual behavior in field crickets. *Behaviour* **17**, 130–223.
2. ALEXANDER, R. D. (1974). The evolution of social behavior. *Ann. Rev. Ecol. Syst.* **5**, 325–83.
3. ALLEE, W. C. (undated). *The social life of animals*. Heinemann, London.
4. ALVAREZ, F., DE REYNA, A., and SEGURA, H. (in press). Experimental brood-parasitism of the magpie (*Pica pica*). *Anim. Behav.*
5. ARDREY, R. (1970). *The social contract*. Collins, London.
6. BASTOCK, M. (1967). *Courtship: a zoological study*. Heinemann, London.
7. BERTRAM, B. C. R. (1976). Kin selection in lions and in evolution. In *Growing Points in Ethology* (eds. P. P. G. Bateson & R. A. Hinde). Cambridge University Press. pp. 281–301
8. BODMER, W. F. (1970). The evolutionary significance of recombination in prokaryotes. *Symp. Soc. General Microbiol.* **20**, 279–94.
9. BROADBENT, D. E. (1961). *Behaviour*. Eyre and Spottiswoode, London.
10. BURGESS, J. W. (1976). Social spiders. *Sci -Amer.* **234**, (3), 101–6.
11. CAIRNS-SMITH, A. G. (1971). *The life puzzle*. Oliver and Boyd, Edinburgh.
12. CAVALLI-SFORZA, L. L. (1971). Similarities and dissimilarities of sociocultural and biological evolution. In *Mathematics in the archaeological and historical sciences* (eds. F. R. Hodson *et al.*). University Press, Edinburgh.
13. CHARNOV, E. L. and KREBS, J. R. (1975). The evolution of alarm calls: altruism or manipulation? *Amer. Nat.* **109**, 107–12.
14. CLOAK, F. T. Jr. (1975). Is a cultural ethology possible? *Hum. Ecol.* **3**, 161–82.
15. CULLEN, J. M. (1972). Some principles of animal communication. In *Nonverbal communication* (ed. R. A. Hinde). Cambridge University Press. pp. 101–22.
16. CULLEN, J. M. Unpublished study on theory of non-genetic evolution.
17. DARWIN, C. R. (1859). *The origin of species*. John Murray, London.
18. DAWKINS, R. and CARLISLE, T. R. (1976). Parental investment, mate desertion and a fallacy. *Nature* **262**, 131–2.
19. DOBZHANSKY, T. (1962). *Mankind evolving*. Yale University Press, New Haven.
20. EHRLICH, P. R., EHRLICH, A. H., and HOLDREN, J. P. (1973). *Human ecology*. Freeman, San Francisco.
21. EIBL-EIBLESFELDT, I. (1971). *Love and hate*. Methuen, London.
22. ELTON, C. S. (1942). *Voles, mice and lemmings*. Oxford University Press.

23. FISHER, R. A. (1930). *The genetical theory of natural selection.* Clarendon Press, Oxford.

24. GARDNER, B. T. and GARDNER, R. A. (1971). Two-way communication with an infant chimpanzee. In *Behavior of non-human primates* (eds. A. M. Schrier and F. Stollnitz). Academic Press, New York and London.

25. HALDANE, J. B. S. (1955). Population genetics. *New Biology* 18, 34–51.

26. HAMILTON, W. D. (1964). The genetical theory of social behaviour (I and II). *J. Theoret. Biol.* 7, 1–16; 17–32.

27. HAMILTON, W. D. (1967). Extraordinary sex ratios. *Science* 156, 477–88.

28. HAMILTON, W. D. (1971). Geometry for the selfish herd. *J. Theoret. Biol.* 31, 295–311.

29. HAMILTON, W. D. (1972). Altruism and related phenomena, mainly in social insects. *Ann. Rev. Ecol. Syst.* 3, 193–232.

30. HAMILTON, W. D. (1975). Gamblers since life began: barnacles, aphids, elms. *Q. Rev. Biol.* 50, 175–80.

31. HINDE, R. A. (1974). *Biological bases of human social behaviour.* McGraw-Hill, New York.

32. HOYLE, F. and ELLIOT, J. (1962). *A for Andromeda.* Souvenir Press, London.

33. JENKINS, P. F. (in press). Cultural transmission of song patterns and dialect development in a free-living bird population. *Anim. Behav.*

34. KALMUS, H. (1969). Animal behaviour and theories of games and of language. *Anim. Behav.* 17, 607–17.

35. KREBS, J. R. (in press). The significance of song repertoires—the Beau Geste hypothesis. *Anim. Behav.*

36. KRUUK, H. (1972). *The spotted hyena: a study of predation and social behavior.* Chicago University Press.

37. LACK, D. (1954). *The natural regulation of animal numbers.* Clarendon Press, Oxford.

38. LACK, D. (1966). *Population studies of birds.* Clarendon Press, Oxford.

39. LE BOEUF, B. J. (1974). Male male competition and reproductive success in elephant seals. *Amer. Zool.* 14, 163–76.

40. LEWIN, B. (1974) *Gene expression—2.* Wiley, London.

41. LIDICKER, W. Z. Jr. (1965). Comparative study of density regulation in confined populations of four species of rodents. *Researches on population ecology* 7, (27), 57–72.

42. LORENZ, K. Z. (1966). *On aggression.* Methuen, London.

43. LORENZ, K. Z. (1966). *Evolution and modification of behavior.* Methuen, London.

44. LURIA, S. E. (1973). *Life—the unfinished experiment.* Souvenir Press, London.

45. MACARTHUR, R. H. (1965). Ecological consequences of natural selection. In *Theoretical and mathematical biology* (eds. T. H. Waterman and H. J. Morowitz). Blaisdell, New York. pp. 388–97.

46. McFARLAND, D. J. (1971). *Feedback mechanisms in animal behaviour.* Academic Press, London.
47. MARLER, P. R. (1959). Developments in the study of animal communication. In *Darwin's biological work* (ed. P. R. Bell). Cambridge University Press. pp. 150–206.
48. MAYNARD SMITH, J. (1972). Game theory and the evolution of fighting. In J. Maynard Smith *On evolution.* Edinburgh University Press.
49. MAYNARD SMITH, J. (1974). The theory of games and the evolution of animal conflict. *J. Theoret. Biol.* **47,** 209–21.
50. MAYNARD SMITH, J. (1975). *The theory of evolution.* Penguin, London.
51. MAYNARD SMITH, J. (1976). Sexual selection and the handicap principle. *J. Theoret. Biol.* **57,** 239–42.
52. MAYNARD SMITH, J. (1976). Evolution and the theory of games. *Amer. Sci.* **64,** 41–5.
53. MAYNARD SMITH, J. and PARKER, G. A. (1976). The logic of asymmetric contests. *Anim. Behav.* **24,** 159–75.
54. MAYNARD SMITH, J. and PRICE, G. R. (1973). The logic of animal conflicts. *Nature* **246,** 15–18.
55. MEAD, M. (1950). *Male and female.* Gollancz, London.
56. MEDAWAR, P. B. (1957). *The uniqueness of the individual.* Methuen, London.
57. MONOD, J. L. (1974). On the molecular theory of evolution. In *Problems of scientific revolution* (ed. R. Harré). Clarendon Press, Oxford. pp. 11–24.
58. MONTAGU, A. (1976). *The nature of human aggression.* Oxford University Press, New York.
59. MORRIS, DESMOND (1957). 'Typical Intensity' and its relation to the problem of ritualization. *Behaviour* **11,** 1–21.
60. *Nuffield Biology Teachers Guide IV* (1966). Longmans, London. p. 96.
61. ORGEL, L. E. (1973). *The origins of life.* Chapman and Hall, London.
62. PARKER, G. A., BAKER, R. R., and SMITH, V. G. F. (1972). The origin and evolution of gametic dimorphism and the male–female phenomenon. *J. Theoret. Biol.* **36,** 529–53.
63. PAYNE, R. S. and McVAY, S. (1971). Songs of humpback whales. *Science* **173,** 583–97.
64. POPPER, K. (1974). The rationality of scientific revolutions. In *Problems of scientific revolution* (ed. R. Harré). Clarendon Press, Oxford. pp. 72–101.
65. ROTHENBUHLER, W. C. (1964). Behavior genetics of nest cleaning in honey bees. IV. Responses of F_I and backcross generations to disease-killed brood. *Amer. Zool.* **4,** 111–23.
66. RYDER, R. (1975). *Victims of science.* Davis–Poynter, London.
67. SAGAN, L. (1967). On the origin of mitosing cells. *J. Theoret. Biol.* **14,** 225–74.
68. SHEPPARD, P. M. (1958). *Natural selection and heredity.* Hutchinson, London.
69. SIMPSON, G. G. (1966). The biological nature of man. *Science* **152,** 472–78.

70. SMYTHE, N. (1970). On the existence of 'pursuit invitation' signals in mammals. *Amer. Nat.* **104**, 491–94.
71. TINBERGEN, N. (1953). *Social behaviour in animals.* Methuen, London.
72. TREISMAN, M. and DAWKINS, R. (in press). The cost of meiosis—is there any? *J. Theoret. Biol.*
73. TRIVERS, R. L. (1971). The evolution of reciprocal altruism. *Q. Rev. Biol.* **46**, 35–57.
74. TRIVERS, R. L. (1972). Parental investment and sexual selection. In *Sexual selection and the descent of man* (ed. B. Campbell). Aldine, Chicago.
75. TRIVERS, R. L. (1974). Parent–offspring conflict. *Amer. Zool.* **14**, 249–64.
76. TRIVERS, R. L. and HARE, H. (1976). Haplodiploidy and the evolution of the social insects. *Science* **191**, 249–63.
77. TURNBULL, C. (1972). *The mountain people.* Jonathan Cape, London.

78. WICKLER, W. (1968). *Mimicry.* World University Library, London.
79. WILLIAMS, G. C. (1966). *Adaptation and natural selection.* Princeton University Press, New Jersey.
80. WILLIAMS, G. C. (1975). *Sex and evolution.* Princeton University Press, New Jersey.
81. WILSON, E. O. (1971). *The insect societies.* Harvard University Press.
82. WILSON, E. O. (1975). *Sociobiology: the new synthesis.* Harvard University Press.
83. WYNNE-EDWARDS, V. C. (1962). *Animal dispersion in relation to social behaviour.* Oliver and Boyd, Edinburgh.

84. YOUNG, J. Z. (1975). *The life of mammals.* 2nd edition. Clarendon Press, Oxford.

85. ZAHAVI, A. (1975). Mate selection—a selection for a handicap. *J. Theoret. Biol.* **53**, 205–14.
86. ZAHAVI, A. (in press). Reliability in communication systems and the evolution of altruism. In *Evolutionary Ecology* (ed. B. Stonehouse and C. M. Perrins). Macmillan, London.
87. ZAHAVI, A. Personal communication, quoted by permission.

Index and key to bibliography

I chose not to break the flow of the book with literature citations or footnotes. This index should enable readers to follow up references on particular topics. The numbers in brackets refer to the numbered references in the bibliography. Other numbers refer to pages in the book, as in a normal index. Terms which are often used are not indexed every time they occur, but only in special places such as where they are defined.